Y0-CDB-905

Radio-Electronics.

FROM
"DRAWING BOARD"
TO
Finished Project

Editors of
Radio-Electronics®

TAB BOOKS Inc.
Blue Ridge Summit, PA

FIRST EDITION
FIRST PRINTING

Copyright © 1989 by TAB BOOKS Inc.
Printed in the United States of America

Reproduction or publication of the content in any manner, without express permission of the publisher, is prohibited. No liability is assumed with respect to the use of the information herein.

Library of Congress Cataloging in Publication Data

Radio-electronics : from "Drawing board" to finished project / by the editors of Radio-electronics.
p.   cm.
Collection of articles from the Drawing board columns of Radio-electronics.
Includes index.
ISBN 0-8306-9133-2   ISBN 0-8306-3133-X (pbk.)
1. Radio circuits—Design and construction. 2. Radio—Apparatus and supplies—Design and construction. I. Radio-electronics.
TK6560.R312   1988
621.3841′2—dc19   88-7796
                                                               CIP

TAB BOOKS Inc. offers software for
sale. For information and a catalog,
please contact TAB Software Department,
Blue Ridge Summit, PA 17294-0850.

Questions regarding the content of this book
should be addressed to:

Reader Inquiry Branch
TAB BOOKS Inc.
Blue Ridge Summit, PA 17294-0214

Edited by David M. Gauthier

CONTENTS

INTRODUCTION

One of the most exciting things you can do if you're into electronics is take an idea from a vague notion in the back of your mind to a working reality in front of your eyes. It doesn't matter if you can buy something similar in a store for half the money, a quarter the time, and an eighth the brain damage. Working out the details of the circuit is usually more satisfying than finishing the project.

There are lots of practical circuits and tips in this collection of my "Drawing Board" columns, but that's not what I had in mind when I sat down to write them. The basic idea behind every one of them isn't to explain how a particular circuit works but to take you through the process of designing the circuit in the first place.

Although circuit details differ from project to project, the general approach to the job is always the same. It doesn't matter what you want to build. Every design—from counters to computers, from light dimmers to lasers—starts out life as an idea, gets clarified on paper, and finalized on a breadboard.

First you have to draw up a list of design criteria. What do you want the circuit to do? What do you want to avoid? What things do you want to include? How is it going to be powered? Defining your goals on paper is the best way I know to firm up an idea. There's no way you can design something without a clear understanding of what you want. The only thing you can do with a rough idea is produce a rough circuit.

Break the project into its component parts. This is the block diagram. In simple terms it lets you see what individual pieces go into the final design. This is one of the most important steps in design; big circuits are really just collections of smaller ones. Ten-page schematics start out as ten separate pages of individual circuits. Nothing gets put together until all the pieces are working.

The completed block diagram is the basis for board work. You pick a place to begin and start getting your hands dirty. And if you're lucky—really lucky—there will be some moments where you'll feel your brain stretching.

So clear your mind, clear your bench, and get to work. I hope you enjoy reading this as much as I enjoyed writing it.

Bob Grossblatt
Circuits Editor

SECTION 1

THE DRAWING BOARD

Voltage Regulators and Power Supplies

ONE THING THAT EVERY ELECTRONICS hobbyist who builds or designs his own equipment will eventually have to contend with is a power supply—it doesn't matter whether you're working on a space shuttle or on an electric toothbrush. It's obvious that there are tremendous differences between the power requirements of a rocket ship and those of a toothbrush, but the point is that if you're designing your own equipment, you're going to have to spend some time thinking about what you want your power supply to do.

It's true that most of the things we'll be discussing in this column can be powered by nothing more complicated than a fresh battery and a pair of alligator clips. From the point of view of elegance however, that approach leaves something to be desired. The power supply you use in your designs can do a lot more for you than just supply power. Most notably, the power supply can provide *protection*.

Even the most carefully designed project in the world can blow up the first time power is applied. But a well-designed power supply can go a long way toward saving you from having to repeat the work you've done in the event of a mishap. It can guard against short circuits; it can limit the current and / or voltage; it can offer protection from transient spikes, and so on. In short, it can be an extremely valuable friend when your project is still in the development stage. Let's look at some of the many possibilities.

Series Regulators

There are many different approaches to power-supply design, but this time we're going to see what we can do with the simple series regulators that we're all familiar with. Those are three-terminal devices that are set up internally to provide a fixed output-voltage of a particular polarity.

The 78xx series of positive regulators (and the 79xx series of negative regulators) are usually used by themselves to provide basic voltage regulation in small electronic systems. Like most things though, those IC's can be made to provide as many exotic features as we want, including the ability to handle much more current than their basic rated capacities would seem to indicate.

Just about everyone is familiar with the circuit shown in Fig. 1-1, a basic five-volt regulator. Capacitor C1 is the huge filter capacitor that sits across the output of the rectifier. It's used to smooth out the spikes (ripple) on the line and to "bruteforce" regulate the voltage going into the 7805. Even though the regulator was designed to reject noise, (referred to in the specs as "ripple rejection"), it can only cope with noise that is a certain proportion of the input voltage. Put simply, the bigger the input-voltage fluctuations, the more noise at the output.

Capacitors C2 and C3 are also filter capacitors. They are generally less than one μF and provide the regulator with local help in dealing with transients. If the regulator is physically far away from the large filter capacitor (C1), a voltage, however small, can develop on the line connecting the rectifier and the IC. The job of C2, therefore, is to make sure that those small voltage transients are eliminated before they reach the regulator. That's why C2 is always located as close to the regulator as possible—in some systems it will be soldered right to the regulator pins. Capacitor C3 does the same job at the output of the device.

Capacitor C4 can be called the "surge capacitor" because its job is to take care of the sudden surges that show up on the system + V line during power-up or power-down. The size of those surges, and consequently the size of C4, depends on the current drawn by the system. Typically though, the value of C4 is somewhere between 10 and 100 μF.

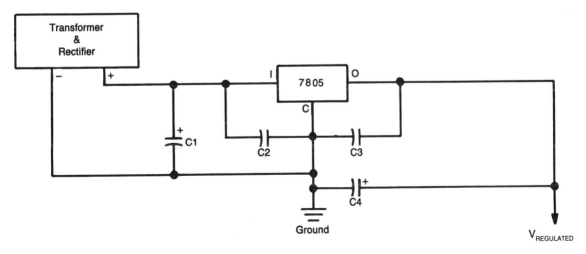

Fig. 1-1.

The 78xx family of regulators (and most other series regulators) is designed to be as foolproof as possible. The regulators monitor their internal temperature; and if they get too hot, they turn off. Short circuits will also cause the IC's to shut down. The trip point isn't a definite figure because it depends on the input / output voltage difference and the temperature. In general, a 78xx-series regulator that is well heat-sinked will be able to handle about an amp—but that's really the upper limit.

Now that we understand the circuit in Fig. 1-1, let's see what's wrong with it. As a side note here, one rule of design is *always* to design with worst-case operation in mind. Remember Murphy's Law and don't forget that one of the drawbacks to original-design work is that the responsibility for backing the warranty is yours.

Problems

Someone once said that there's no such thing as a free lunch, and that applies here. We're using capacitors to help the regulator minimize noise and transients, but capacitors cause another problem. A rapid reduction in either the input or output voltage will cause the capacitors to discharge. How much discharge current is generated depends on a lot of variables—the values of the capacitors, the rate of voltage reduction, and so on. Most regulators are built to withstand a certain amount of discharge current, but the unpredictability of the amount of that current makes for a real problem. In order to put things in perspective, consider that a 10 μF capacitor can develop 20-amp spikes when it's shorted.

If you're designing a power supply only for low-current systems, that doesn't present much of a problem. But if you're going to need a healthy amount of current, something has to be done to protect the regulator against accidental capacitor-discharge.

It will help to think of the regulator as a bunch of control circuits with a beefy pass-transistor at the output. In Fig. 1-2 we show only C1 and C4; since C2 and C3 are relatively small, we don't have to pay as much attention to them.

In the case of an input short, we have a big problem with C4. When the input short occurs, C1 will discharge through it and the input voltage to the collector of internal pass-transistor Q1 will rapidly fall to zero. That means the output voltage will be greater than the input voltage. Since C4 will have stored a nice, healthy charge, it will start to discharge. Some of the discharge current will go though R_{sc}—the equivalent resistance of the regulator's protection circuitry. If the current is substantial enough, R_{sc} isn't going to pass it fast enough and the emitter-base junction of Q1 is going to be reverse biased. If the current is great enough, Q1 is going to break down and the regulator will be—to use a technical term—zapped.

Fortunately, an output short isn't anywhere near as serious. In that case, C4 will discharge across the output short and the input voltage will be greater than the output voltage. Luckily, the regulator was designed to deal with that. It will start to pass more and more current until its thermal-overload point is reached and it shuts down. Remember, the regulator was designed to source current, not sink it. That's why an input short is so much more potentially dangerous than an output short.

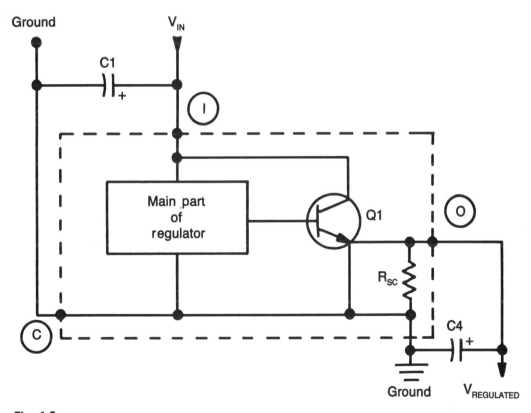

Fig. 1-2.

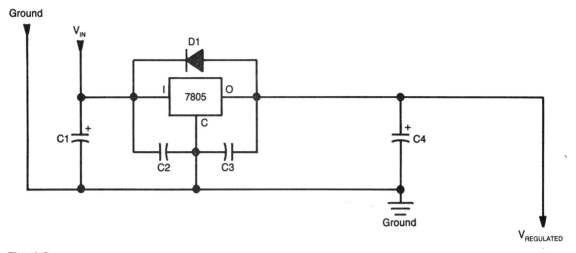

Fig. 1-3.

In order to protect against input shorts, we have to find a way to provide an escape path for the discharge current of C4. We'll add a diode as shown in Fig. 1-3.

Adding protective diode D1 gives us a really slick solution to the problem. If the input shorts out now, the discharge current from C4 will forward bias the diode and all the "bad" current will be shunted to ground through the input short.

Making Fixed-Output Regulators Adjustable

We ended out discussion of voltage regulators looking for the answer to an intriguing problem. We were trying to figure out why diode D1 (see Fig. 1-4) does not conduct even though it seems that current is flowing in the right direction. I'm sure that most of you figured out why D1 doesn't shunt any current when the regulator is operating normally, but let me give you the answer anyway. In normal operation, D1 doesn't conduct because it *can't* conduct. Remember that the regulator's input voltage is always greater than its output voltage. Regulation is like anything else in life—you don't get something for nothing. With the 78xx series of regulators the input has to be at least 2 volts higher than the output—if you want everything to work the way it's supposed to. Therefore, the potential at D1's cathode will be at least 2 volts higher than at its anode. Since that reverse-biases the diode, there won't be any current flow.

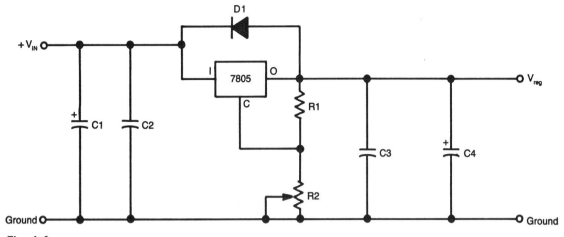

Fig. 1-4.

When things *do* blow up, D1 has to be hefty enough to handle the discharge current from C4, and it has to start working quickly enough to get rid of the current before the regulator is damaged. That's why our choice for D1 has to be a fast-acting silicon diode that is capable of handling the surge current without being destroyed. Depending on the parameters of the circuit, any member of the 1N400x family of diodes is a good choice.

Our regulator circuit so far is fine—if all you need is a fixed standard-voltage and not much in the way of current. But suppose some circuit that you're building requires an oddball voltage, and you're sure that the current requirements are going to be rather large. Obviously we have to take our regulator a bit farther.

Raising the Output Voltage

There are two ways that we can raise the output voltage—let's call them the easy way and the hard way. Figure 1-4 shows the easy way. Resistors R1 and R2 form a voltage divider across the regulated output-voltage. As we move the wiper of R2 toward ground, we change the ground reference of the regulator and trick it into putting out a higher regulated-voltage. I won't bother you with all the grisly details of the math, but the formula is: $V_{reg} = 5 + (5 / (R - 1) + I_{sb})R2$ where I_{sb} is the standby current used by the regulator. (For a 7805 with no load at its output, I_{sb} is usually about 8 mA.) Obviously that figure will change as the circuit is put under load. The formula shows us exactly what we're doing—we're adding the voltage generated across the voltage divider to the normal output of the regulator. That approach to an adjustable regulator is okay for small currents, but it leaves a lot to be desired when we're looking for substantial amounts of current and real flexibility. As you can see from the formula, the lowest voltage we can get with that arrangement is the basic voltage of the regulator. Not only that, but we're seriously interfering with the stability of the output voltage.

The reason for the instability is that the regulator can only handle a fixed amount of power. Now, since we all know that power is the product of the voltage and the current, the more voltage we get, the less current we can safely draw from the regulator. As the voltage at the IC's internal pass transistor increases, the internal protection of the IC automatically reduces the short-circuit trip-point. Not only that, but if you use that circuit and the current drawn from it comes close to the trip point, R1 and R2 will start to get warm and change value. And, changing those values will also change the output voltage. Fortunately the IC is protected and thermal runaway isn't possible, as it is with transistors. Even so, it's annoying, to say the least, to have your power supply drop out every time you put a moderate demand on it.

A better, but slightly more complicated, way to make an adjustable supply from a fixed regulator is shown in Fig. 1-5. I've left out the capacitors to make the drawing clearer and

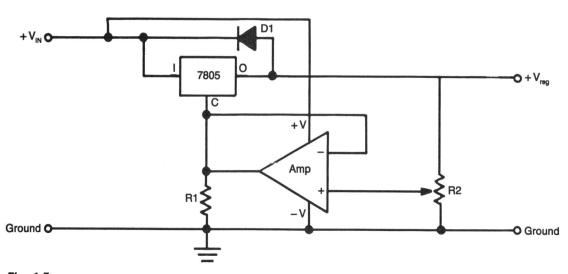

Fig. 1-5.

cleaner. If you compare this circuit with the one shown in Fig. 1-4, you'll see that an op-amp has replaced the potentiometer. Since some of the newer op-amps have input impedances as high as a trillion ohms, it's pretty safe to say that the output of the regulator won't be loaded down as it was in the first circuit. The op-amp is set up as a voltage follower, which means that it's nothing more than a buffer—a noninverting amplifier that's used to isolate one part of a circuit from another.

As we move the wiper of R2 away from ground, an increasingly greater voltage is present at the ground terminal of the regulator. That raises its ground reference and tricks it into putting out a higher voltage. The drawback is that the minimum output voltage is still going to be the standard output of the regulator, in this case five volts, plus the voltage drop across R1, which is about two volts. On the other hand, we can get the regulator to put out as much as 20 volts without worrying about limiting the current-handling capability of the IC, or degrading the voltage regulation when the numbers start to get large. This adjustment range of 15 volts or so is just about the most you can hope to safely get out of a fixed regulator like those in the 78xx series. You can move the entire output range slightly by changing the value of R1, but trying to extend the range too far will cause you to run the risk of doing severe damage to the IC. After all, remember that it *was* designed to be a fixed regulator.

Lower Output Voltages

There is a way to change the range of the (fixed) adjustable regulator so that we can drop its output voltage much closer to ground. The way to do that is to allow the op-amp to swing its output below ground. Figure 1-6 shows how that can be done. By using a center-tapped transformer we can reference the ground terminal of the regulator and the $-V$ input of the op-amp to a level below system ground. Then, the regulator will reference its output voltage to a point below ground. We have to be careful when we set this sort of thing up however, because the regulator's internal pass transistor was designed to source current, not to sink it. If the output level of the regulator gets below ground we're going to have the same problem we had before and the same thing will result—one french-fried regulator.

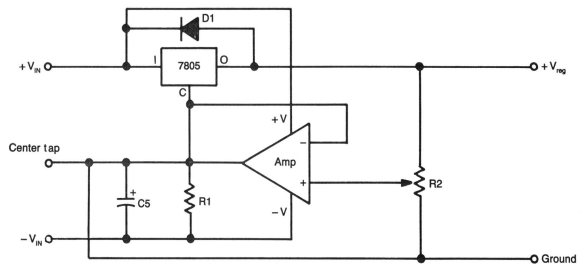

Fig. 1-6.

Another alternative is to build a small circuit that generates a true negative supply from the positive system-voltage. We don't need a lot of current since the op-amp draws next to nothing and the regulator is only using the negative voltage as a reference point.

You can always use a different regulator, such as the LM317, which was designed to go as low as 1.25 volts and up to more than 35 volts. Your regulator-circuit design will be easier, but you won't learn anything. Besides, there's a certain amount of perverse pleasure that comes from making an IC do something it wasn't expected to do in the first place.

One other thing you should remember is that there's always at least a two-volt drop across the regulator. If you're planning on designing a power supply that can put out 20 volts, make sure that you have at least 22 volts available at the input to the regulator. The same goes for the amount of current you can draw—you can't get out of the regulator what the transformer and rectifier can't put in.

Increasing Current-Handling Capability of Regulators

The trend in modern logic families is to make them operate with smaller and smaller amounts of power. (I suppose the ultimate goal is the family that can run on potential energy!)

Lower power-requirements get rid of the necessity for wrist-thick cables and glass insulators, but there's an even more important benefit. Lower power means smaller, and less complicated, regulator circuits. Some IC's even have the regulator circuitry built onto the chip's substrate. Less current-draw means that the layout of the $+V$ run on printed-circuit boards is much simpler. Remember that when heavy amounts of current are running through a trace on a board, a potentially troublesome voltage drop will be generated because of the resistance (however small) of the copper trace. That can lead to inductive oscillation and other nightmares.

That "low power" side benefit, however, can tend to make you a bit forgetful when you're developing a power supply. LED's, relays, and other things can still gobble up current at an alarming rate. A power supply that can deliver half an amp may seem perfectly adequate for, say, a CMOS circuit—and it is. Unfortunately, when we start asking the circuit to turn something on or light something up, the current draw is going to increase dramatically and our half-amp supply is rapidly going to drop dead.

The voltage-regulator circuit that we've been developing can so far safely supply about a half amp over its full range, but it's a smart move to design it so that it can provide a lot more. Since the internal circuitry of the 7805 is limited to less than one amp, it's obvious that we're going to need some other device to provide the additional current.

Adding a Pass Transistor

In Fig. 1-7, we've added a transistor and a resistor to take care of the additional current. For simplicity's sake I haven't drawn in the rest of the circuit we've developed so far. All the current that goes into the regulator has to pass through R_B since it's in series with the regulator input. Ohm's law tells us that as the current flow through a resistor increases, so does the voltage developed across it. The base-emitter junction of Q1, a pnp transistor, is in parallel with R_B. As long as the current flowing through the resistor is below a certain level, just about all that's going to happen is that the resistor will get a little warm. At some point, however, a voltage drop across R_B is going to get high enough to turn on the transistor, which will start to pass current through its collector. That current is added to the current supplied by the regulator and

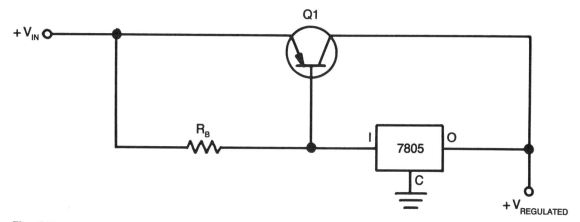

Fig. 1-7.

allows the draw on our power supply to be increased by the amount that the Q1 can handle without blowing up.

Transistor Q1, then, is used as a switch that senses when the regulator output is near some limit and turns on to provide the extra current that the regulator can't handle. The turn-on point of Q1 is determined by the value of R_B and the base-emitter voltage of Q1. One other thing to be aware of is that the difference between the input and output voltages is going to change. Since Q1 and R_B are in series with the regulator input, the voltage drop across them has to be added to the inherent 2-volt drop of the regulator. That is important to remember when we're figuring out how much voltage we need at the output of the rectifier.

Short-Circuit Protection

Before we start doing any arithmetic to calculate the value of R_B we have to add some short-circuit protection to the circuit. I know you're thinking that we took care of that earlier, but we've now added active components to the input. If the output is shorted now, all our earlier protection springs into action—but it only takes care of the regulator. The collector of Q1 is going to be shorted out and the transistor is going to start passing current through the short. It will rapidly exceed its maximum collector-current rating, and all you'll be able to do is administer the last rites.

That is, to say the least, an undesirable state of affairs. In Fig. 1-8 we've added a safety net for Q1 in the form of Q2 and R_S. Those of you with sharp eyes will recognize that

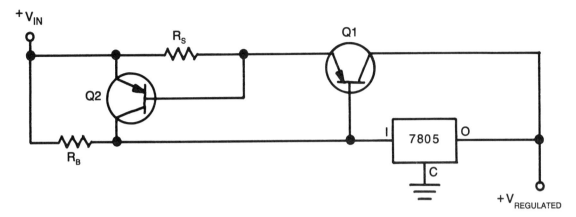

Fig. 1-8.

those two new components form a switch in exactly the same manner as R_B and Q1. The same sort of analysis also applies.

All the current that flows through Q1 has to pass through R_S. When a certain point is reached, the emitter-base junction of Q2 is going to conduct and the transistor will turn on. When it does, it will lower the voltage across R_B and turn Q1 off. Since Q2 isn't going to turn on until the power supply is providing really large amounts of current, we need a hefty transistor there. It has to handle pretty close to the sum of the short-circuit currents of both the 7805 and Q1.

Since there are more components connected in the circuit between the base and emitter of Q1, the math needed to calculate the values of the two resistors is going to be more complicated. Rather than going through it however, let's make a few intelligent assumptions and see if we can make life easier.

If we use silicon transistors for Q1 and Q2, we know that the base-emitter voltage is going to be about .65 volts when the transistor is turned on. As long as the voltage is below that, the transistor will be turned off.

Now let's look at Fig. 1-7 again and assume that Q1 isn't there. The 7805 needs about 8 mA to operate—the rest of the current it passes is available to whatever circuit it's powering. The regulator can handle half an amp without any problem, but let's be on the safe side and arrange for Q1 to turn on when the regulator draw exceeds 250 mA. Since the turn-on voltage for the transistor is 0.65 volts, calculating the value of R_B is a snap: $R_B = E / I = .65 / .250 = 2.6$ ohms.

Now, it's true that the emitter-based junction of Q1 is in parallel with R_B so that bunch of arithmetic isn't strictly correct. Remember, though, that the apparent resistance of the junction when the transistor is in cutoff is pretty high. It's not really accurate to talk about the resistance of a transistor (or any semiconductor, for that matter), because they're dynamic devices and we should more properly refer to their "impedance." That's the DC resistance coupled with an AC component. For our "real world" circuit, however, the difference doesn't amount to much and we can ignore it.

If you look at Fig. 1-8, you'll see that we have to go a little farther in figuring the value of R_B. Since both R_B and R_S are across the emitter-base junction of Q1, both their values have to be taken into account when we figure the trip point of Q1. Once again, the "resistance" of Q2 in cutoff is high enough for us to ignore it and just work with the resistor values.

Since R_S has to pass all the current that flows through Q1, we have to decide what we're going to let the maximum current be. Five amps is a good value for our regulator circuit—more than that will cause design problems we don't want to get involved with. Just as was the case with Q1, Q2 will start conducting when its emitter-base voltage reaches 0.65 volts. If we want that to happen when Q1 is passing 5 amps, R_S has to be on the order of 0.13 ohms. The total resistance we need to turn on Q1 is 2.6 ohms. Since R_S must be .13 ohms, the new value of R_B will be 2.47 ohms.

Now, I'm the first to admit that those are pretty oddball values for resistors. You can't exactly amble down to your local resistor store and buy a 2.47 ohm resistor. There are ways around that, though.

Finishing Up About Regulators

As you're probably aware by now, there's a lot more to consider when you're building a power supply than how to get a battery out of a blister pack. Over the last few months we've gone through the design considerations necessary to build a voltage regulator complete with all sorts of bells and whistles. Before we wrap up this project, be aware that we've just barely scratched the surface. There are lots of different sorts of power supplies and we've been looking at only one particular kind. There are design engineers who do nothing else in their professional lives but design power supplies. (That may strike you as being a bit boring but it certainly gives you an idea of the size of the subject.)

Where's the Surprise?

Figure 1-9 is the full schematic of the regulator circuit we've designed. The values for all the components are shown, and if you look closely you'll see that we've added something to the short-circuit protection we discussed. We've put in a switch, S1, and broken R_s into two parts. The reason that was done is the surprise I promised.

I told you that there was a way around trying to locate a 0.13 ohm resistor. You probably realized that we would par-

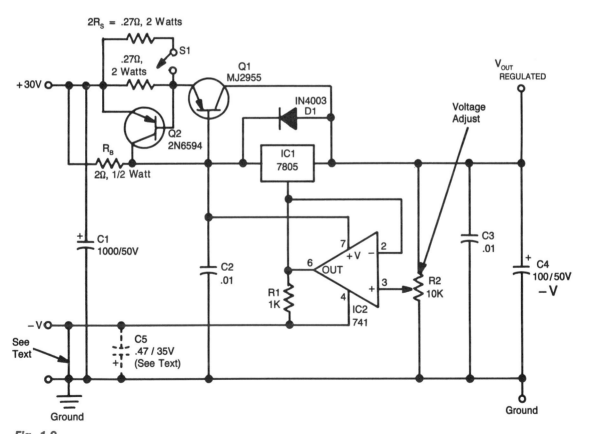

Fig. 1-9.

allel a few resistors to get the value we wanted and that's exactly what I did. When we close S1 we put the two resistors in parallel and arrive at a value of 0.135 ohms—nothing really surprising about that. Opening S1 puts only one of the resistors in the line and lets us change the trip point of Q2 to two amps. We can now switch-select the short circuit trip-point to be either 2.4 amps or 4.8 amps. The real treat comes when we do the arithmetic necessary to calculate the wattage we need for the resistors.

Let's assume that we have the full value for R_S. We saw that it takes a 0.65 volt drop across the emitter-base junction of Q2 to make it conduct. The short circuit trip-point would be: $I = E / R = .65 / .135 = 4.82$ amps.

That's pretty close to our original target of 5 amps. Now let's calculate the wattage we need for R_S: $P = I^2R = (4.82)^2(.135) = 3.14$ watts.

In the interest of safety, and with proper respect for Murphy's law, let's call it 4 watts. This makes R_S a pretty hefty resistor, and probably not too easy to find. It is easy however, to find a 0.27-ohm, 2-watt resistor, so we'll put two of those in parallel. Since the wattage adds when resistors are in parallel (do you know why?) we have 0.135 ohms at 4 watts when S1 is closed and 0.27 ohms at 2 watts when S1 is open. If you want to verify the calculations for the circuit with S1 open, follow the format I've just used. It's satisfying to see the numbers work out so neatly.

More About Component Values

If you remember last month's discussion of R_B, you'll recall that we had to do some math to arrive at a value of 2.47 ohms. A look at Fig. 1-9 will show you that R_B is now specified to be 2 ohms at half a watt. Changing the value to 2 ohms from 2.47 ohms was done in the interests in reality. The only place you'll be likely to find a 2.47-ohm resistor is the same place you'll find a 0.13-ohm resistor—nowhere. You may be considering the idea of using a trimmer potentiometer in place of a fixed resistor. Don't do it—it's a bad idea.

There are two no-no's as far as R_S and R_B are concerned—trimmers are a bad idea, and so are wire-wound resistors. The reasons for avoiding those things have already been discussed and you should realize what they are. Go back through our discussion of regulators so far and figure out the answer.

Table 1-1.

R_S	$R_S + R_B$	I_{RB} at Q1 turn-on	P_{RB}
0.135Ω	2.135Ω	305 mA	0.185 watts
.27Ω	2.27Ω	286 mA	0.163 watts

The calculation of the wattage I'm going to leave as an exercise for you. To be fair, though, I'll give you the answers. They are shown in Table 1-1. Make the same assumptions we did about the transistor "resistance" and see if you can work out those values.

As you can see, calling for a half watt for R_B is a bit of overkill, since the calculated value never even reaches one-quarter watt. When we started this exercise in design I told you that one of the cardinal rules of design was always to aim for worst-case operation. By specifying a half-watt resistor we're adding a safety factor of more than two. If you can get your hands on a one-watt resistor of the right value it wouldn't hurt to use that either—you can never have too many watts.

The transistors listed for Q1 and Q2 have the collector-current maximums the circuit calls for. Since they're being used as switches we don't have to worry about any of the parameters we'd have to consider in an analog circuit (frequency compensation, gain, and so on). It's really just a matter of finding some silicon transistors of the right polarity that can handle the calculated current. The MJ2955 can handle 15 amps at its collector, and our circuit only calls for 5 amps—the safety factor again.

Q2 has to be able to stand the combined short-circuit currents of both the regulator and Q1—that means a collector current of 0.035 + 4.82 = 5.13 amps. A 2N6594 can handle 12 amps. I called for those transistors because they're easy to find. If you want to substitute other transistors, that's okay—just watch the collector ratings, and make sure you use silicon transistors. If you want to use germanium ones, all the calculations we've done will have to be redone. You tell me why.

Other Components

I've called for a 741 as the op-amp to be used for IC2. Actually you can use just about any op-amp there. The 741 seemed to me to be a good choice because it's cheap (always an important design consideration), available, and has internal current-limiting and frequency compensation. If you have some other op-amp lying around, use it. The higher the input impedance the better. Since it's only a buffer, the requirements aren't at all critical.

The last thing to look at is C5. It handles the transients from the negative supply, if you use one, to lower the range of the circuit following the guidelines we developed. It goes without saying that if you go this route the connection shown in Fig. 1-9 between the $-V$ input and ground should be ignored—so I won't say it.

There's nothing sacred about the choice of IC1. If your voltage requirements are going to be consistently higher than five volts use one of the other regulators in the series—the 7812 or 7815 for example. Try giving the op-amp a bit of gain by putting some resistance on the feedback line from pin 6 to pin 2. The point of these columns is for you to learn—and there are only two ways to do that. Read everything you can get your hands on, and get your hands in everything you read. Components are cheap enough so that blowing a few up is still less expensive than going to school . . . and more instructive as well.

You should have learned enough to design your own negative supply. Try it and let me see what you come up with. Just be careful. You're playing around with circuitry that can deliver a lot of power. If you heat-sink everything—and you should—you're looking at circuitry that can melt the tip of a screwdriver!

Successful Designing and Breadboarding

Just about everyone is familiar with Zeno's story about Achilles' race with the tortoise. But for those of you out there who *don't* remember the story, it goes something like this: Achilles and a tortoise decided to have a race. Since Achilles was the Jesse Owens of his day and the tortoise even slower than an apology, they gave Achilles a handicap by letting the tortoise have a fifty-foot head start. (It was only fifty feet because even ancient bookies didn't like to take chances.) On the day of the race, the tortoise moved up fifty feet, Achilles laced up his sneakers, the gun went off, and so did Achilles.

Here's the twist: Every time Achilles covered half the distance to the tortoise, he still had half the remaining distance to go. And, according to Zeno, since time and space are infinitely divisible, Achilles never even passed the tortoise, much less got to the finish line. What Zeno didn't tell us was that even though Achilles never passed the tortoise, when he got close to him, he caught him, cooked him, and ate him.

Believe it or not, there's a valuable lesson to be learned from that story for all of us who like to design our own equipment. Culinary tastes aside, the moral of the story is really that even though you may not be able to get there, you can get close enough for all practical purposes—never mind what the mathematicians tell you!

Paper and The Real World

In a nutshell (which is where all valuable lessons are to be found), there's a big difference between brainwork and boardwork. Things that work out perfectly well on paper have a nasty habit of blowing up when you build them. Sometimes it's because you overlooked something in designing them, and sometimes it's because you messed up in building them, but most of the time it's just because we live in the real world. Formulas, graphs, and tables of values stretched out to endless significant figures can lull you into thinking that the same sort of precision exists here in the real world. It doesn't.

Not only do unexpected things crop up when you start breadboarding, but even before that. The final paperwork often turns out to be the result of compromises and approximations. The difference between the words "standard value" and "calculated value" is what keeps the variable-resistor and -capacitor manufacturers prosperous and smiling. I challenge you to take any design problem and solve it using only standard-value components. That's only one example of how things on paper differ from things in the real world.

One of the magic moments in life comes when power is first applied to the first breadboard version of an original circuit design. Since it never works the first time around, the first thing to do, obviously, is to check the breadboard against the paperwork. Let's assume everything checks out. Now, the problem can only be in one of two places: the physical devices or the original design. I'm going to show you how you can cut the amount of your troubleshooting by as much as fifty percent, and maybe even more.

Using the Right Parts

Creating a completely workable electronic system is a difficult job any way you look at it—no matter what you want it to do. Even a fairly simple design-goal can involve some rather complicated circuit parameters. What you should be aiming for is a method that makes your life as simple as possible. One way to do that is to make sure that the components you're using in the breadboard circuit are all 100% functional.

Anyone who breadboards circuits using "junkbox" parts is asking for trouble. Those components are available all over the place—unmarked IC's ("you test 'em . . . we don't have the time") and surplus boards loaded with everything imaginable except an explanation. As far as the former is concerned, if their time is too valuable why should yours be any less so? And surplus boards hardly ever have socketed IC's. Removing one of those non-socketed packages safely is something like trying to fix a lightbulb. I suppose it's possible, but since you can get a new one for thirty-nine cents, why bother? Anyway, the chances are that you're going to destroy the IC as you remove it.

In a circuit-design that calls for loads of IC's I don't think you'll save more than a few dollars by using surplus ones. Considering the price of IC's today, salvaging surplus ones isn't worth the bother unless you put a really low value on your time. I don't mean the time it will take to find and remove the components, either. You're going to spend a lot of time troubleshooting your circuit when it doesn't work. And that's to say nothing of the damage a bad IC can do to the more expensive parts of your circuit. Blowing up a twenty-dollar memory because you used a bad surplus twenty-cent gate is, to say the least, ridiculous.

One of the least discussed linear qualities of digital IC's is their degree of "goodness." Although they basically operate with only half and low states, they can be 100-percent functional, pretty good, OK, pretty bad, or any degree in between. A surplus IC that tests OK may blow up in your circuit. That's because your testing probably won't make the exact demands on the IC that your circuit will. Internal diodes may be marginal, pass transistors may be degraded, and so on.

Another problem with surplus parts: A manufacturer may have made a seemingly standard part with different specifications to meet the particular requirements of a customer. For example, Radio Shack used what looked like a quad OR gate in the early version of the TRS-80. One of the gates wasn't used—because the chip had only three functional gates on it!

As I've said before, when you design your own equipment, you're the one responsible for the warranty. The key to making your bench time as productive as possible is to eliminate any possible source of error. Breadboarding with anything other than brand-new components is the best way I know to grow gray hair. As far as prototyping is concerned, junkbox components are just that—junk.

Data Books

Successful original design is the result of three things: a clear bench, a clear mind, and a clear understanding. The first you get with a broom, the second with a brain, and the third with a stamp (your brain will help here too). Major component-manufacturers publish data sheets, application notes, and data books. Those not only provide you with the operating parameters of the various devices, but they are loaded with suggestions and hints that can make your design work a lot easier.

Once you've decided what components you're likely to need in your design, find out who makes them and drop them a note requesting information. Chances are you'll get back a listing of the data sheets and application notes they have available. There's usually a nominal charge for those things, but the information that you'll get will save you all sorts of bench time. The application notes are particularly useful because they can give you ideas about how to begin your design. More often than not they'll suggest an approach to the problem that never occurred to you.

In Table 1-2 I've given you a list of some of the major manufacturers who publish data books. It's by no means an exhaustive list, but it's a good beginning.

Table 1-2.

American Microsystems 3800 Homestead Rd. Santa Clara, CA 95051	**Motorola Semiconductor** Box 20912 Phoenix, AZ 85036
Exar Integrated Systems, Inc. 750 Palomar Avenue, PO Box 62229 Sunnyvale, Ca 94088	**National Semiconductor** 2900 Semiconductor Drive Santa Clara, CA 95051
Fairchild Semiconductor 313 Fairchild Avenue Mountainview, CA 94040	**RCA** Box 3200 Somerville, NJ 08876
General Instrument Corporation 600 W. John Street Hicksville, NY 11802	**Signetics** 811 East Arques Avenue Sunnyvale, CA 94086
Gould Batteries 931 Vandalia St. Paul, MN 55114	**Solid State Scientific** Montgomeryville Industrial Park Montgomeryville, PA 18936
Intel Corporation 3065 Bowers Avenue Santa Clara, CA 95051	**Sprague Electric Company** Semiconductor Division 115 Northeast Cutoff Worchester, MA 01606
Intersil, Inc. 10710 N. Tantau Avenue Cupertino, CA 95014	**Texas Instruments** Box 5012 Dallas, TX 75222

Working with Counters

IC-fabrication technology has come a long way since the first IC rolled off the production line a mere twenty or so years ago. Component density has gone from four transistors on the early chips to over four hundred thousand transistors on current ones. These mind boggling numbers have led to all sorts of good things—from five-dollar microprocessors to blister-packed digital watches sold next to the canned soup in the supermarket. The result of all this on someone (like me) who occasionally likes to re-invent the wheel to solve circuit problems has been quite extraordinary.

I've had to re-define the wheel.

What is new, expensive, and exotic today is most definitely cheap and ho-hum tomorrow. I can remember using loads of power-gobbling gates and flip-flops to build counters. Today that approach to a circuit design would be ridiculous because the array of features in available MSI (*Medium Scale Integration*) counters can take care of any design problem you can imagine. Counters have to be considered a basic building block of digital design—in other words, a one IC addition to a circuit.

Now, the word "counter" takes in a lot of territory—anything that does first one thing and then another in a pre-arranged sequence can be called a counter. Just about the only thing they have in common is that they need a power supply and some sort of clock. There are lots of ways you could divide them up but since we're calling them a basic building block, we'll make a basic two divisions—counters with a one-and-only-one type of output and those with encoded outputs.

Every logic family has its own array of counters and for our purposes, anything we say about the counters in one family will be more or less true of the counters in any other family. We'll restrict our discussion to CMOS counters since we're more interested in finding out how to use them than in chopping the top off the package and looking at the silicon.

The 4017 is a good example of a counter that has only one output decoded at a time. It has ten outputs and they go high one at a time in fixed sequence as long as the ENABLE and RESET pins are held at ground. A high on the ENABLE pin will disable the clock input and the counter will ignore incoming clock pulses. A high on the RESET pin will make the "0" output go high; it will stay that way until the RESET pin is grounded again. There's also a "carry" output that divides the input clock by ten—it's high for counts zero through four and low for counts five through nine. This IC is really a shift register with a few added bells and whistles. There are, however, some interesting things we can learn from it and some extremely useful things that it can do when we put it to work for us.

First of all, this is a synchronous counter. That means that all the internal flip-flops are triggered by the incoming clock at the same time. The other possible arrangement is called a ripple counter, meaning that the internal clocking takes place like a row of dominoes—each stage triggers the next stage. Ripple counters are cheaper to make; but they're much slower than synchronous ones since stage changes happen in serial, rather than parallel, fashion. They will also temporarily output incorrect counts while the dominoes are still falling. That glitchy period is euphemistically called the "settling time" but it would be more accurate to call it the time when the output of the counter was just plain wrong. Since the speed at which CMOS operates is a function of, among other things, the supply voltage, lower voltages can lead to delays many microseconds long. During those microseconds the counter output is not exactly something you'd want to take to the bank.

The one-and-only-one type of counter can come in really handy when you have to solve certain design problems. The keyboard data encoder we designed showed two of the many possible uses for this type of counter. We used it there to select a particular switch at the keyboard and also as a sequencer to control the order in which data was latched onto the bus. That is, of course, by no means all it's good for.

The best way to understand how the IC is used is, naturally enough, to actually use it. Since the 4017 has outputs that sequence one after another, probably the most basic circuit we can build is the sequencer shown in Fig. 1-10. We're using one half of a 4011 to make a simple clock we can use to drive the 4017. Any other oscillator would be just as good. The frequency of the 4011 clock follows the form F = 1 / 1.4RC. Since we want to be able to see the 4017 outputs in action, we'll pick values for the clock components that slow it down enough for us to watch things happen. The values shown will give a clock frequency of about 3 Hz—a nice compromise between visibility and impatience.

Everything else in the circuit is straightforward. By tying both the ENABLE and RESET pins to ground, the 4017 will count from zero to nine over and over again. Now, that isn't the most exciting thing I've ever seen but even this circuit has some important real-world uses. What you're looking at is a one-IC method of delaying clock pulses by a time period exactly equal to N clock pulses. All you have to do is route your clock to the input of the 4017 and pick off whichever phase-shifted output you want. Of course your input clock will have to be running ten times faster than the frequency you want to see at the output, but that's not much of a problem.

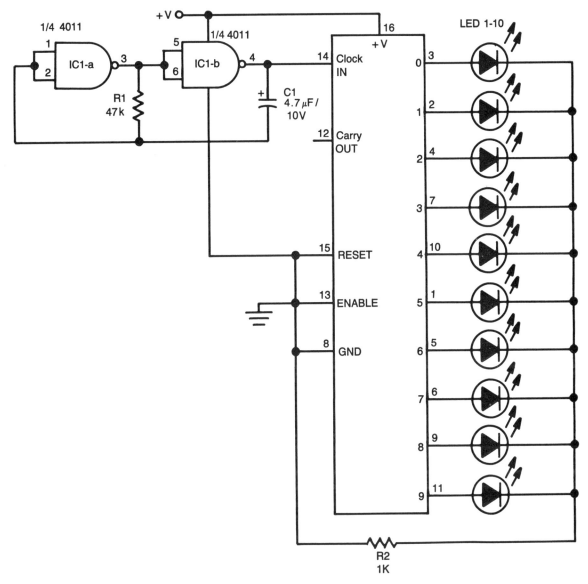

Fig. 1-10.

We can spice things up even more by using the ENABLE and RESET pins. Tying the ENABLE pin to a particular output means that the 4017 will count to a certain number and then stop. Doing the same thing with the RESET pin will give you a really down-and-dirty method of frequency division. Since the IC will reset to zero whenever the selected output goes high, any of the chip's outputs in sequence before the selected one will go high at a rate equal to f / N where f is the input clock frequency and N is the number you're dividing by.

Someone once said that there's no such thing as a free lunch and that applies here as well as anywhere. While it's obviously true that you can divide a clock down this way, it's also unfortunately true that you're paying a price for simplicity. First, the duty cycle of the output will be something like $1 / N$. This makes sense because the outputs go high for one full cycle of the input clock and remain low for the rest of the time. I said "something like $1 / N$" because there's a certain amount of uncertainty that's caused by the weirdness that goes on when the selected output goes high and the IC resets. That leads to the second price we have to pay.

When you operate the IC at 5 volts, the propagation delay (the time it takes for the IC to change to a new state) from one output to the next is about 500 nanoseconds. This means that there will be a 500-nanosecond delay between the time an input pulse is detected and the IC puts a high on the next output in sequence. Let's assume we have the RESET pin tied to output 4 and output 3 is high. Along comes the next input clock pulse—it's detected and the internal machinery of the IC starts to decode it. When that operation is finished it simultaneously turns off output 3 and turns on output 4 (remember that this is a synchronous counter). So far so good.

When output 4 goes high, it brings the RESET pin high and causes the IC to turn off output 4 and put a high on pin 3—the first output in its sequence. The problem crops up because the 4017 features asynchronous reset. That means that reset takes place whenever the RESET pin is brought high. In an IC with synchronous reset, the reset operation wouldn't happen until the next clock pulse arrived at the input. The 4017 is counting as a synchronous counter but reset is happening in a ripple fashion. Our problem is that the IC ignores incoming clock pulses when the RESET pin is high as well as during the entire reset operation. A quick look at Table 1-3—which shows us the characteristic operational times for a 4017 operating from 5 volts—illustrates exactly what the problem is.

In the best of all possible worlds, therefore, there's a built-in period of almost 1 microsecond (500 + 450 nanoseconds) during which the 4017 is performing its reset operation. We have to wait for the selected output to be decoded and then

Table 1-3.

Operation	Propagation Delay	Pulse Width	Transition Time
Decoded output	500 nanoseconds	200 nanoseconds	300 nanoseconds
Reset	450 nanoseconds	200 nanoseconds	250 nanoseconds

twiddle our thumbs while the reset operation is carried out. Since the clock input is disabled half this entire time (during reset), we'd better make sure that no clock pulses show up at the input because they're going to be ignored. The price, therefore, that we're paying for down-and-dirty frequency division is a cutback in the maximum input frequency we can have and the possibility of glitches in the count.

Solving the Reset Problem

The most important thing we learned from the discussion of counters in general and the 4017 in particular is this: you get what you pay for. Using the 4017 to get nice cheap frequency division was like a lot of things in life—it seemed like a good idea at the time but when we put the IC to work we only got half of what we expected. The circuit was certainly cheap enough but the results were a far cry from nice. Two major problems showed up that really limited the usefulness of the circuit. The first, asynchronous reset, introduced some unpredictablity in the output. The second problem was that the output duty cycle would change as we changed the number we were dividing by. Now, for some applications those may not be important but for others, they can be a real problem.

Not only that, but it's a good rule of thumb in design to limit the times you're willing to shrug your shoulders and compromise. After all, one of the main reasons you're designing something from scratch is to have the circuit do exactly what you want it to do. There are already more than enough times in life when you have to meet something halfway.

The Reset Problem

Let's tackle the reset problem first. In a nutshell, asynchronous reset means that the IC will reset itself whenever the reset pin is brought high. Not only is the operation independent of the input clock but you also have no control of the time the RESET pin remains high. Fortunately, that problem can be licked with a little bit of imaginative gating.

The trick is adding synchronous reset to the 4017 is being able to control the RESET pin. We need some sort of gating arrangement that will make it go high when we want; and more important, make it go low when we want. We also need some way of making sure that latter signal comes when RESET is completed. Remember that the 4017 is disabled as long as the RESET pin is held high.

Let's digress for a bit. What we're talking about here is a gating arrangement that has two independent inputs and whose output will change state when triggers are applied to each of the inputs. So, as you've probably guessed by now, we need a flip-flop. Now, you can use some standard-type flip-flop such as the 4043 or 4044; but in line with the established traditions of this column, let's see what we can do with a bunch of simple gates.

Figure 1-11 shows how we could build a simple flip-flop. It's made from two inverters and is mechanically triggered. If

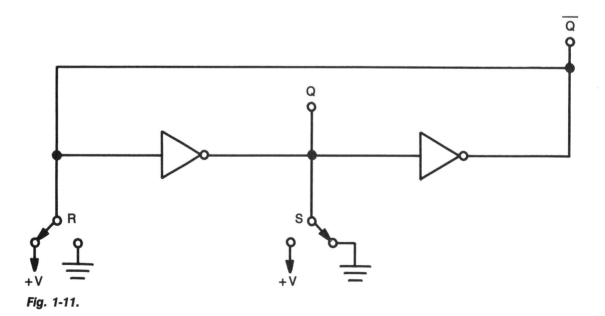

Fig. 1-11.

we connect the R switch to ground the Q output will be positive and the $\overline{Q}$ output will be negative. In fact, triggering either of the outputs will produce entirely predictable results and obviously stable outputs. The only thing we can't do is connect both the R and S inputs to the same polarity—if we did, the whole thing would obviously go up in smoke. Since we all know that if something bad can happen it *will* happen, we'd better look further for our needed flip-flop.

In Fig. 1-12, we've done the same sort of thing with a pair of NAND gates. The operation of the circuit is more interesting and, ultimately, more useful. Let's assume the R input is connected to +V and see what happens as we switch the S input between +V and ground.

If we connect the Q input to ground, the $\overline{Q}$ output will be high because a NAND gate has a high output if one or more of its inputs are grounded. Since the $\overline{Q}$ output is also connected to one of the legs of the first NAND gate we have two highs there and the $\overline{Q}$ output will be low. If we switch the S input to +V, the output of the second NAND gate won't change because the other leg is still being held low. Suppose we now connect the R input to ground while the S input is switched to +V. With one leg grounded, the $\overline{Q}$ output will go high and the two highs at the inputs of the second NAND gate will make the $\overline{Q}$ output low. As you could predict, that flip-flop only responds to negative triggering because of the basic operation of the NAND gate.

The only thing to watch out for when you're using this flip-flop is to make sure that the S and R inputs aren't grounded at the same time. If that does happen, both outputs will be high and the circuit will be unstable. In practice, the last input to

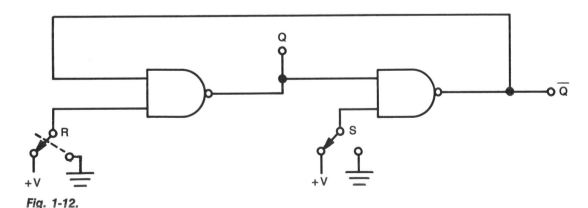

Fig. 1-12.

be grounded will decide the ultimate state of the flip-flop. But don't believe us—build it and try it yourself.

We can build the same sort of circuit with NOR gates and the flip-flop will respond to positive triggering. Use the previous

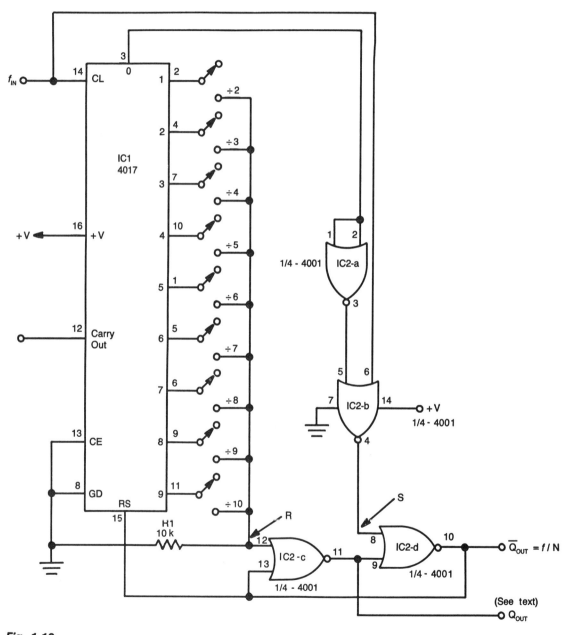

Fig. 1-13.

Table 1-4.

Flip-Flop Truth Table			
R	S	Q	$\overline{Q}$
LOW	LOW	No Change	
LOW	HIGH	HIGH	LOW
HIGH	LOW	LOW	HIGH
HIGH	HIGH	Not Allowed*	
*See text			

discussion as a guide and trace through the operation of the NOR gate flip-flop so you understand how it works.

Now let's get back to our original problem. The circuit in Fig. 1-13 uses a NOR gate flip-flop to control the operation of the reset pin on the 4017. We're using NOR gates because they respond to positive triggering and the outputs of the 4017 are active high. The truth table for the flip-flop is shown in Table 1-4. The not-allowed state with NOR gates is having both the flip-flop inputs high. This isn't a problem, since the internal gating of the 4017 guarantees that only one output can be high at any one time.

As long as none of the switches are closed, R1 holds the input of the flip-flop low. That means that the $\overline{Q}$ output will be low regardless of what is happening at the S input. The reset pin of the 4017 is also held low and the chip is enabled. Now let's close one of the switches and see how the circuit works.

When the selected output goes high, the R input of the flip-flop goes high and a high appears at the $\overline{Q}$ output. This resets the 4017, the selected output goes low, and the 0 output, pin 3, goes high. Remember that the 4017 outputs go high in turn and whatever output you select will be low immediately following a reset pulse. That makes the R input low and control of the flip-flop moves over to the S input. You can see from the truth table in Table 1-4 that we need a positive pulse there to make the flip-flop change state and put a low at the $\overline{Q}$ output to release the 4017's RESET pin.

The Ø output of the 4017 is inverted by NOR gate IC2-a and presents a low to one leg of NOR gate IC2-b. The other leg of the gate is connected to the input clock and when a low appears there, IC2-b goes high and resets the flip-flop. That releases the RESET pin and enables the 4017. If you keep the switch closed at the keyboard, the circuit will reset over and over at the same point. The result will be a series of pulses at the $\overline{Q}$ output equal to the input frequency divided by whatever number you chose.

The circuit gives the 4017 a reset operation that is both synchronous and locked to the input clock. By following everything carefully you should have no trouble understanding how we did it. Remember that the reset operation starts on the positive half of the incoming clock cycle and is ended on the negative half of the same clock cycle. Since the input clock will be running faster than the pulses at any of the 4017 outputs, we don't have to worry about glitching in the count.

A side advantage of that approach is that the Q output of the flip-flop will give us an output wave that is equal in frequency to the $\overline{Q}$ output but opposite in sign, that can come in handy for some things and is especially nice since we're getting it for free.

If you want to cascade several 4017's together to increase the range of division, you won't be able to use the carry output, pin 12. Since that pin is high for the first half of the 4017's full count and low for the second half, frequency division of less than six will mean that the carry pin never goes low. The 0 output will, however, always go through a full cycle no matter what division you're doing, so you can take your signal from there.

The Duty Cycle Problem

Now that we've solved the reset problem, let's look at the duty-cycle problem. In case you forgot what it is, we discovered that the duty cycle of the output would change every time we divided by a different number. It would follow the form 1 / N where N is the number you're dividing by. More specifically, the high time would be equal to the period of the incoming clock, and the low time would be equal to N − 1 times the period.

If you're dividing by an even number, some simple gating would let you get an output with a nice 50% duty cycle but trying to do the same thing with an odd number would be—well, odd.

One of the basic rules of design is that there's a better way to do everything and that's true here. When simple problems generate overly complex solutions, it's time to scrap your whole approach and start over with a different color paper. In this case, squaring up the duty cycle not only calls for a different approach, it calls for a different IC—a different kind of counter.

Generating Sinewaves with the 4018

By this time we should all be familiar with the unbreakable first rule of electronic design: *Brainwork before board work*. If you can't get it down clearly on paper, you can't design it, much less build it. (I believe that there's some sort of natural law that governs the relationship between the weight of the finished product and the paperwork it generates. If anyone knows what it is, please let us know!) Paperwork always chops mind-boggling design problems down to a manageable size and also lets you concentrate all your energy on specific design problems.

We spent a little time breaking down the problem of using the 4018 to generate sinewaves. Although that's certainly not the most complex problem you'll ever see, it is important to remember that the design approach that you take is as important as the design itself. As a matter of fact, the initial approach will more often than not shape the final product.

Generating Sinewaves

Take a look at the output waveforms of the 4018 shown in Fig. 1-14. It puts the procedure to follow (and the problem it causes) in black and white for us to look at. And it should give you some idea as to how to go about using that IC. As you can see, the 4018 provides phase-shifted outputs that are delayed by exactly one incoming clock pulse. Not only that, but we've already seen that the output duty cycle is nice and square. If we sum the outputs together properly, we can produce a digital waveform that can be filtered to any degree of smoothness desired by the circuitry that's tied to its output.

If the outputs (Q_1 through Q_5) of the 4018 are added together using equal value resistors, we're going to wind up with the very familiar and entirely predictable waveform shown in Fig. 1-15a. If you squint your eyes and imagine the waveform to be all smoothed out you'll see that the best that we can hope to get from the circuit in Fig. 1-15b is a triangular wave.

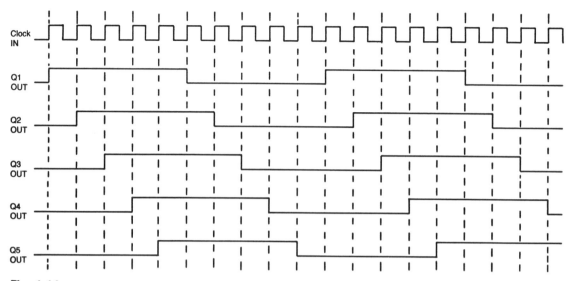

Fig. 1-14.

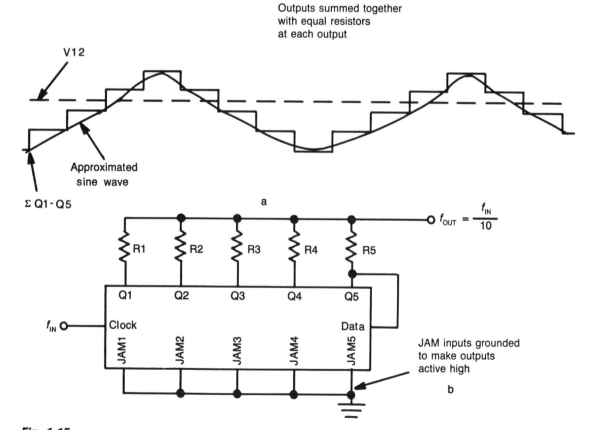

Outputs summed together
with equal resistors
at each output

Fig. 1-15.

Obviously, our approach is on the right track but the problem is a little more complex than it first appeared. While, it is evident that we have to add the IC's outputs together, it should also be evident that we have to give more thought to how we do it. The shape of the wave that's generated by the 4018 depends on the values chosen for the summing resistors. Determining the values of those resistors, however, is something else. There's no way to avoid doing some math; but let's see if there's some way to at least cut the required calculations down to a slightly less formidable size. Once again we have some paperwork to do.

Now, as everybody knows, there are lots of different ways to go about solving a problem. Which one you pick depends on the problem, but remember that the idea behind all of them is to cut down the amount of work you have to do. Let's attack our problem with the most basic approach—common sense!

In Fig. 1-16 we see a composite output waveform from Fig. 1-15 and we have also overlaid it with an approximation of the sinewave that we're trying to generate. Certain things should become clear almost immediately.

At the sinewave approaches its maximum positive and negative values it flattens out. The staircase shape that was generated using equal value resistors has sharp peaks at those points and therefore, doesn't really fit the curve. That simple observation leads us to a sledgehammer-type fix. All we have to do now is to lose the output of the 4018 that's causing those peaks. In practical terms that means getting rid of the Q_5 output. As you can see from Fig. 1-15b, we're using that output for two purposes: It's one of the data outputs and since it's the last output in the chain, we're connecting it back to the data input of the IC. Remember that we have to make sure

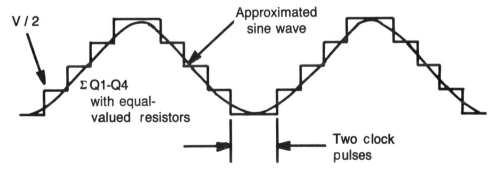

Fig. 1-16.

that the incoming data at the clock input is constantly recirculated around and around the daisy-chained flip-flops in the 4018. Any change in the input data to the IC has to be fed into it at the clock input and not the data input. All that we're using the data input for is to make sure that whatever we feed into the 4018 stays there.

Losing the Q_5 output means that we don't use it to recirculate the data. The result of eliminating that output is shown in Fig. 1-17: There have been two changes, the one we expected and one that's just one of those lucky breaks. The most obvious change is the flattening at the top of the waveform. That's what we expected and isn't really any great surprise. We've overlaid the output waveform with a sinewave again and you can see that it is a better fit than we got earlier in Fig. 1-15b. Even though the drawing of the sinewave is crude, you can see that it's going to be a much better fit.

The second change isn't quite as obvious because the drawings aren't exactly to scale. Since we've lost one of the

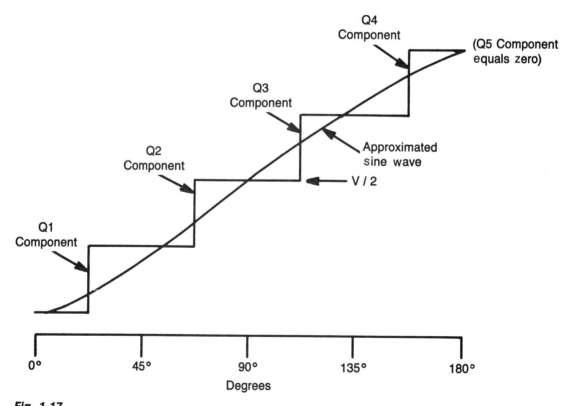

Fig. 1-17.

36

outputs (the one that produced the top of the output waveform) the top only remains for two incoming clock periods instead of three, as it did in Fig. 1-15b. That not only helps us fit the crest of the sinewave but gives the rise and fall on either side a better shape making the fit even better. Of course, as we've seen over and over again, you can't get something for nothing and we're paying a price here as well. Let's not forget that while we may have an easier time fitting the curve, we've lost one of the outputs and consequently our resolution has suffered. But, as with so many other things, trade-offs are the name of the game in electronics as well.

Picking the ideal resistor values to give us the best approximation of a sinewave involves a lot of math. The principle behind the whole thing, however, isn't really that hard to visualize. Figure 1-18 gives us a graphic representation of the problem. What we're looking at there is the first 180 degrees of the sinewave. The generating of the sinewave means that all of the outputs are going to come into play during each half of the full-cycle. Take another look at Fig. 1-14, you'll see that the sequential rise and fall of the flip-flop outputs determine the shape of the summed waveform. Because each of the outputs is out of phase (or delayed) by exactly one

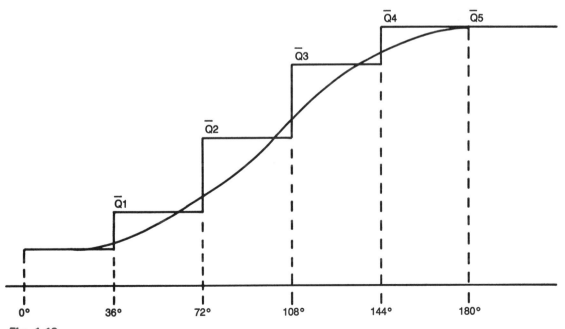

Fig. 1-18.

incoming clock pulse, each of the outputs controls the amplitude of the output waveform at 45 degree (or 180 / 4) intervals. (We're dividing by four instead of five because the Q_5 output is not being used. Even though we're allowing for the time it takes to change state, it adds nothing to the amplitude of the output waveform.)

Finding the correct resistor values, therefore, means a bit of trigonometry and some more analysis. Don't be put off by the math; it's not all that difficult and understanding it only involves common sense and curiosity—two very important tools for anyone who wants to be involved in electronics.

Smooth Out the Sinewave Generator Output

Anyone who get interested in electronics and gets really hooked on it will progress through a number of clearly recognizable steps. In the beginning, you buy light-dimmer kits and burn your eyes out trying to read obscure directions written in muddy print. The next step is to buy components and, armed with a chart that lists the resistor color code and a soldering iron, burn your components up trying to build a light dimmer of your own design. Somewhere around here you begin to understand that there's more to electronics than Ohm's law, and you begin to read.

Now, we're all familiar with the truth of Grossblatt's 12th Law: He who doesn't have his head in a book has his head in something else. But the more general the rule, the more exceptions there are to it, and that applies here as well. After you've plowed through enough abstracts and journals, you'll learn how to apply Grossblatt's 27th law: What is written on paper is not carved in stone.

The difference between theory and practice is the difference between brain damage and common sense. The difficult task of plowing through countless reams of paperwork filled with obscure equations can often be eliminated by taking a look at the original problem on a different-color or walking away and letting your subconscious take over.

The perfect example of that is the problem facing us at the moment—finding the resistor values for our digital sinewave generator. There are three ways to go about finding the answer. 1) Trial and error. 2) Mind-warping math. 3) Common sense.

The first one is OK, but only gives answers for a particular application. The second is OK for people who wear a bathing suit with shoes and socks. That leaves us with the third.

Believe me when I tell you that the standard method for calculating the resistor values involves math so hairy . . . well, even with a lot of equipment it would be difficult. The Fourier transforms and Fibonacci numbers are the easy part. The hard parts can only be solved using a variable interossiter. (Do any of you remember what that is or know how to spell it?)

Getting Around the Math

But seriously folks, the math is both complicated and unnecessary. We can get within several decimal places of the calculated values by using common sense and a bit of elementary arithmetic. Let's take a good look at the problem. Figure 1-19 shows the circuit we're going to use; Fig. 1-18 shows 180 degrees of the waveform that we want, and a couple of helpful hints.

You'll remember that we're not using the Q_5 output of the 4018 because it's a quick and dirty way to make the waveform conform more closely to a sinewave. The resistors on the remaining outputs will determine the shape of the wave we generate but—and this is important—we still have to allow for the time used by the Q_5 output. In other words, no matter how many 4018 outputs we decide to use, it's still going to take 5 incoming clock-pulses to make the 4018 repeat itself. That means any calculations that we do have to take into

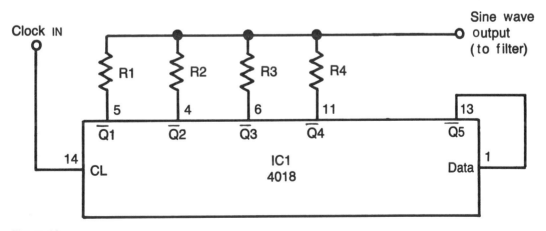

Fig. 1-19.

account the fact that there will be 5 discrete 4018 output states for each 180 degrees of the sinewave.

In practical terms, each incoming clock pulse will control 36 degrees (180 / 5) of the sinewave. Q_1 will determine the amplitude of the sinewave 36° into the cycle, Q_2 will determine the amplitude of the sinewave at 72°, and so on until we get to Q_5. Even though we're not using it, we still have to allow for the time it takes for the 4018 to cycle through it. Make sure you understand that!

Translating that bit of common sense to actual resistor values is really simple. We look up the sine of the angles we're interested in and generate a table like that shown in Table 1-5. We already know the angles we want—they're listed together in the appropriate column in the table. The last column translates that data into something that's easier for us human beings to use. All that we've done is to make the relative proportions a bit more evident by dividing all the sine values into .951.

So, you may well ask, what do we have to do next?

Well, believe it or not, that's all we have to do! All our work is done and the only arithmetic (as opposed to mathematics) we have left is some multiplication. What the last column in the table is telling us is that in order to generate a sinewave using 4 of the outputs from 5 daisy-chained flip-flops, the resistor values have to be in the proportions indicated. Pick a convenient value for R_2 and R_3, do the arithmetic, and you've got your resistor values! Of course, you might have a hard time finding standard-value resistors in the right ratios but that's a common problem—and, naturally enough, it has a common solution. You can use precision resistors if you're rich enough, or trimmers if you're not. In any event, we've got it made!

Table 1-5.

Output	Angle	Sine	Normalized Value
$\overline{Q1}$	36°	0.588	1.617
$\overline{Q2}$	72°	0.951	1.000
$\overline{Q3}$	108°	0.951	1.000
$\overline{Q4}$	144°	0.588	1.617
$\overline{Q5}$		Not Used	

40

I know you haven't seen the math we managed to avoid doing, so you can't appreciate the kind of work we saved. What we've done is a classic example of how a common-sense approach to a problem can eliminate a lot of effort and keep the men in the white coats from your door. Let's go through the reasoning behind all that and make sure we understand it.

If the data is recirculated in the 4018, 5 incoming clock cycles have to pass before the output states start to repeat. One complete cycle of the 4018's outputs will be needed for each half of the sinewave we want to generate, regardless of how many of the outputs we actually use. That means that each incoming clock pulse will come when the sinewave we want to generate has advanced *one fifth of half its full cycle* or 36 degrees (180 / 5).

The amplitude of the sinewave at any point on the curve can be found by looking up the sine of the angle. Once we've listed all the ones we need, we can find the ratios of the resistors we need to generate the wave. See that? It *is* simple!

If you decide you want to use more flip-flops in the sinewave generator that you build, you'll have to recalculate the resistor values. Just go through the same reasoning we outlined and you won't have any problems.

From a practical point of view, I would recommend that you standardize the lowest value resistor at 10K or so and use trimmers to get the other values that you'll need. Just measure the 10K resistors to get the *exact* value, and do the arithmetic to find out where to set the trimmers. Set them out of the circuit and use a bit of nail polish to lock them in place before you put them on the board.

There are other parts to this sinewave generator we're slowly designing: the input clock, frequency selector, and the output filter. The most interesting one is the input clock. With a little bit of thought, we can make it variable so that the frequency selector can be something as simple as a potentiometer. That *was* one of our original design criteria.

Since we've already seen that the input clock has to run ten times faster than the maximum sinewave frequency we want to produce (remember—the 4018 is set up to divide by ten), we need a clock that can be tuned over a 1000:1 range with a twist of the wrist. There are a couple of things that come to mind that will fill the bill, but we ought to think about refinements such as crystal control of the frequency, stability, low-power requirements, and all those other good things.

Finishing the Sinewave Generator

This may seem hard to believe, but we've just about reached the end of our discussion on digital sinewave generators. We're going to put a lid on the whole thing and take a look at the bits and pieces that have to be added to make a working unit. If you have been following along you know that all that's left to do is to select an oscillator for the front end and a filter for the back end. Let's take care of the rest of the design and then talk a little bit about the advantages that come your way by generating sinewaves with digital, rather than analog circuitry.

If you were to single out the most often designed (or redesigned) circuit in the history of electronics, the oscillator would win hands down. The reason for mentioning that is to point out that it would be a waste of brain power to sit down and design a special one for our application.

Our requirements are sufficiently simple to make use of most existing general-purpose oscillator circuits. All we need is something that is easily tunable over a 1000:1 range, has good stability, is foolproof, and has a garbage-free output. The bells and whistles would include an adjustable pulse-width, minimum parts count, low power consumption, and so on. But first we must select an oscillator!

Oscillator Circuit

Obviously, the first thing that should come to mind is the venerable 555 timer. Not only is it easy to use, but the design equations are simple to apply. And if power requirements are really a problem, we can use the CMOS version (7555). Both parts also satisfy another unspoken design criteria. That is, they are common enough to be found almost everywhere except in a box of *Cracker Jacks*.

Figure 1-20 shows the basic layout for an oscillator using a 555 timer. We are looking at one IC and four components—not a whole lot of parts. Now, there is absolutely no doubt (in

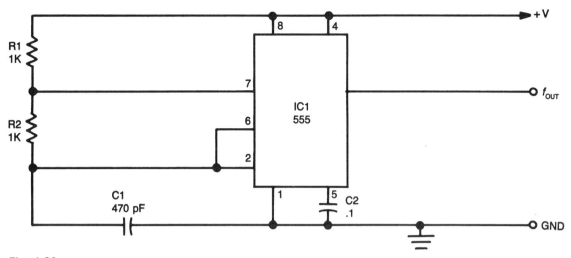

Fig. 1-20.

no doubt (in my mind) that all of you out there know everything there is to know about the 555; but just to make sure, let's go through the circuit, equations, and the math.

The first thing that must be realized is that the output duty-cycle will not be exactly 50%. With the circuit in Fig. 1-20, we can get extremely close to 50%, but never quite reach it. The reason for that capacitor C1 charges up through resistors R1 and R2, but discharges only through R2. If you were to look inside a 555, you would see that there's a transistor handling the discharge half of the cycle and that its connector is collected to pin 7 of the IC. Its emitter is connected to ground and the base sits on the output of an internal flip-flop.

When the output of the 555 is high, capacitor C1 charges up until the voltage across it reaches about $\frac{2}{3}$ of the supply. At that point the internal transistor turns on providing a discharge path for capacitor C1 through resistor R2, while the flip-flop changes states and the IC's output goes low. When the capacitor voltage gets down to about $\frac{1}{3}$ of the supply, the flip-flop again changes states and the transistor turns off. Capacitor C1 starts to charge up again through both R1 and R2.

Now that we understand what determines the duty cycle of the basic 555 oscillator, a quick look at some of the design equations for the IC will show you what has to be done to square up the duty cycle. Because we're looking at a series R-C timing circuit, the equations related to the duty cycle will be as follows:

$$\text{high time} = K(R1 + R2)C1$$
$$\text{low time} = K(R2)C1$$

The "K" in the formula is a constant, which we'll get to in a minute. But for now, we can ignore it because when we calculate the duty cycle, we'll see that "K" takes care of itself and disappears.

$$\begin{aligned}\text{duty cycle} &= \text{high time} / \text{low time}\\ &= K(R1 + R2)C1 / K(R2)C1\\ &= (R1 + R2) / R2\end{aligned}$$

The duty cycle is controlled by the value of the two resistors (R1 and R2), so we can approximate 50% by making R1 really small when compared to R2. A ratio of up to two or three thousand to one is perfectly workable. The frequency of the oscillator, as we all know, is just the reciprocal of the period (or one divided by the period). Rather than just throw formulas at you, let's briefly reason the whole thing out.

$$\begin{aligned}\text{period} &= \text{high time} + \text{low time}\\ &= K(R1 + 2R2)C1\end{aligned}$$

The frequency (f) is therefore:

$$f = 1 / K(R1 + 2R2)C1$$

For the circuit shown in Fig. 1-20, the value of 'K' is about .695. That value comes about because the 555's internal comparator is set to trigger at ⅔ of the supply voltage. The actual math to derive that value is far from being interesting but, if you want to, you can check it out in any good data book that includes the 555 timer. You can set the trigger level higher by applying a dc voltage to pin 5, but for our application it's not needed. Besides, we can bypass it with C2 to help make the circuit a bit more immune to noise. Now, let's move on to the sinewave generator.

Generator Circuit

The original design criteria we set up for the sinewave generator called for an output range of 100 Hz to 100 kHz. Since the 4018 is divided by ten, the range of the oscillator must be 1 kHz to 1 MHz. Because the upper limit of the 555 is just around 1 MHz, we should be all right.

Figure 1-21 is the layout for the complete sinewave generator with all the values shown. You can plug your own numbers into the formulas we've just described and change anything you want. Nothing is engraved in stone and, to tell the truth, you'll learn much more if you experiment with different value components.

Potentiometer R3 serves as the frequency selector, making the 555 tunable from about 1 kHz to 1 MHz. I say "about" because it's impossible to find standard value parts that will plug into a formula and give you exactly the results you want.

Resistors R4 through R7 are the famous ones we spent so much time examining. They determine the shape of the wave you're going to get from the 4018. As shown, the fixed resistors are set at 10,000 ohms; and trimmers were used for the $\overline{Q}_1$ and $\overline{Q}_4$ outputs, because the value we want for each

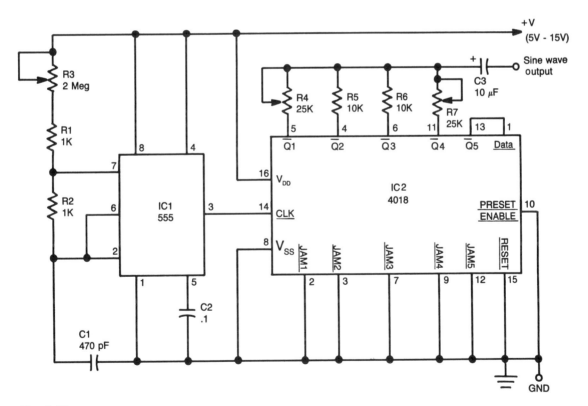

Fig. 1-21.

of them is calculated to be 1.617K—not exactly a standard value!

Not only that, but the strange resistor value is correct only if the fixed resistors are exactly 10,000 ohms. Since we all routinely use 5%- or 10%-tolerance resistors, the best thing to do is take two 10,000-ohms units, measure them (out of circuit), and then do the arithmetic necessary to find the value for the two trimmers. After you set the trimmers, it might be a good idea to put a drop of lacquer (or nail polish) on the turnscrews to hold them in place.

The output filter is just a 10-μF capacitor. It should work well enough to smooth the wave out for most applications. However, if you feel that it does not fulfill your needs, by all means substitute some other type filter. Harmonics are not going to be much of a problem because they're quite a bit above the fundamental frequency and they are a good deal smaller in amplitude.

You may wonder what's so great about generating sinewaves digitally. Well, there are several advantages: good stability, low-power requirements, no complicated R-C setups, and so on. As far as I'm concerned though, the biggest advantage is that you can easily and reliably generate frequencies (of any precision) even well below 1 Hz. Since the input clock is running ten times higher than the output of the 4018, you can produce frequencies down around .0001 Hz. Try doing that with capacitors!

The output of the whole generator can be fed into anything you want from a simple buffer to an amplifier: What you use depends on what you want to do. An op-amp would be great to use as a buffer / amplifier here. Also, we could make the gain adjustable, and so on. But wait a minute: Op-amps need a bipolar supply and all that is available in our circuit is a single-sided one!

That's quite a problem—or is it?

Section 2

NEW IDEAS

Simple Tesla Coil

THERE IS ONE IMPORTANT THING to keep in mind before we even begin: The Tesla coil described here can generate 25,000 volts so, even though the output current is low, *be very careful*!

The main component of the Tesla-coil circuit is a flyback transformer. You can get one from a discarded TV.

The first thing you must do is to get rid of any excess wire or other debris that's on the transformer's core, as shown in Fig. 2-1. Leave the high-voltage winding alone; but if there is a capacitor at the end, it should be removed.

After that, you can start winding a new primary coil. Begin by winding 5 turns of No. 18 wire on the core. Then twist a loop in the wire and finish by winding five more turns. Wrap with electrical tape, but leave the loop exposed.

A four-turn winding has to be wound over the ten-turn winding that you've just finished. That is done the same way. First wind two turns of No. 18 wire, then make a loop, and finish up by winding two more turns. Again, wrap the new winding with electrical tape, leaving the loop exposed.

When the windings are finished, the two loops shouldn't be more than ¼-inch apart (but take care that they do not

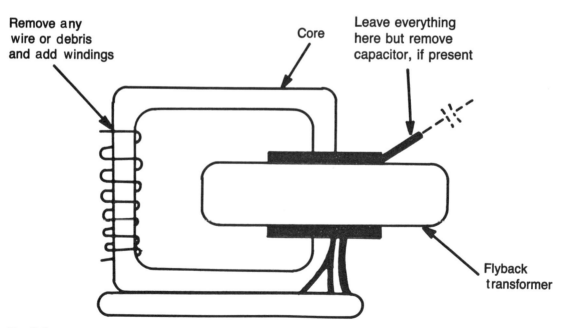

Remove any wire or debris and add windings

Core

Leave everything here but remove capacitor, if present

Flyback transformer

Fig. 2-1.

48

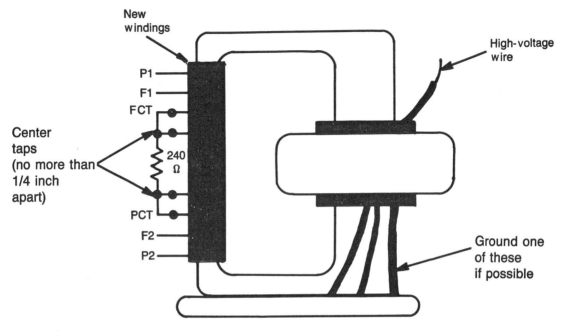

New windings

P1
F1
FCT

Center taps (no more than 1/4 inch apart)

240 Ω

PCT
F2
P2

High-voltage wire

Ground one of these if possible

Fig. 2-2.

touch). Connect a 240-ohm resistor between the two loops. The modified transformer now should look like the one shown in Fig. 2-2.

Connect the transformer as shown in Fig. 2-3. The 27-ohm resistor and the two transistors should be mounted on a heat sink and *must be insulated from it*.

The output of the high-voltage winding should begin to oscillate as soon as the circuit is connected to a 12-volt dc power supply. If it does not, reverse the connections to the base leads of the transistors. In normal operation, you should be able to draw 1-inch sparks from the high-voltage lead using an **insulated** screwdriver.

Automobile Ignition Substitute

Here's a device that can help you find out what's wrong when suddenly one morning your car refuses to start. The ignition substitute described here can even be used to verify that repair work done to your engine by someone else has been done correctly.

Basically, the ignition substitute provides a constant power-source for the ignition coil. Its frequency (0.5-1.0 kHz)

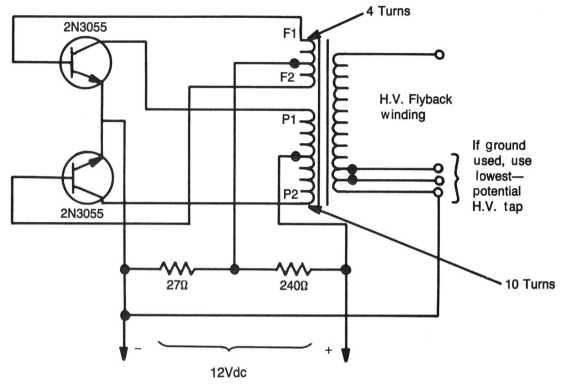

2N3055

2N3055

F1 · 4 Turns

F2

P1

P2 · 10 Turns

H.V. Flyback winding

If ground used, use lowest— potential H.V. tap

27Ω   240Ω

−   +

12Vdc

Fig. 2-3.

is that used by an 8-cylinder engine with an idling speed of 650 RPM, and the unit provides a rapid spark at a 17% duty cycle, while nonetheless staying within the power-dissipation limits of the components.

Construction is straightforward, and any method can be used. The circuit, shown in Fig. 2-4, consists of a 555 timer IC configured as an astable free-running multivibrator that is used to drive a high-current npn transistor, such as a 2N6384. (That transistor should be heavily heat-sinked because it may be drawing several amps over quite a long period of time.)

The coil ballast can be from 0.68 to 6.5 ohms, depending on what's available. The 2.5-ohm, 20-watt ballast shown in Fig. 2-4 works well. All the other resistors can be either ¼- or ½-watt devices, and the capacitor between pins 1 and 5 of the 555 can range from 0.01 to 0.05 μF. Do not omit the 100-volt, 0.05 μF capacitor across the transistor; it prevents voltage spikes from damaging the device. Use 4-foot-long clip leads to obtain power directly from the automobile's battery; that length is suggested for convenience.

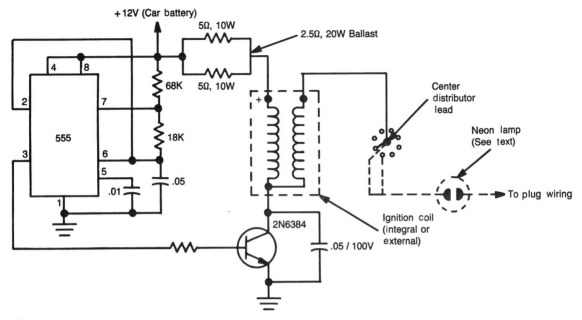

Fig. 2-4.

You can use either your car's own ignition coil, or a separate one. If you choose the latter, be sure to disconnect the one in the vehicle. A good coil will produce a spark between the high-tension lead and ground about ¼- to ½-inch long, and a strong bright spark across a plug with a gap smaller than 0.040-inch. That, by the way, is the first test for ignition problems.

To determine whether there's a problem with the car's distributor, supply a spark derived from the ignition substitute to the center distributor lead leading to the rotor and slowly rotate the distributor cap. Crank the engine and, at some point, the engine should catch and run. If the engine cannot be started, but seems to be trying to, the problem is probably in the timing chain, valves, camshaft, or elsewhere. If the engine doesn't even try to start, inspect the rotor, cap, wires, and plugs for damage. Once the ignition problem has been found and corrected, the normal procedure for setting the timing and dwell should be followed.

Do not attempt to adjust the distributor using the ignition substitute. That cannot be done because the spark the substitute produces is slightly different from that produced under normal conditions.

Although designed for an 8-cylinder engine, this device can be used with other types. In addition, a neon bulb can be added to the circuit to verify the presence of a spark, and, in fact, can be used as a timing light if placed close to timing marks that have been painted white with fingernail polish.

Automobile Locator

Have you ever had trouble finding your car in a crowded parking lot? If so, here's a device that will be of some help.

This automobile locater is made up of two parts. The first is an RF oscillator, whose circuit is shown in Fig. 2-5. The second is a sensitive receiver; that circuit is shown in Fig. 2-6.

The heart of the oscillator is a 555 timer IC. Its frequency—just below the AM broadcast-band—is determined by R1, R2, and C1. A tank circuit (C2 and L1) is used to tune the transmitter. The antenna is coupled to the transmitter through C3. Since efficiency is not very important here (output power should be kept under 100 mW), the length of the antenna can be kept short. A telescopic antenna or a length of hookup wire will work quite well. The only thing that is important is that the antenna be vertically polarized.

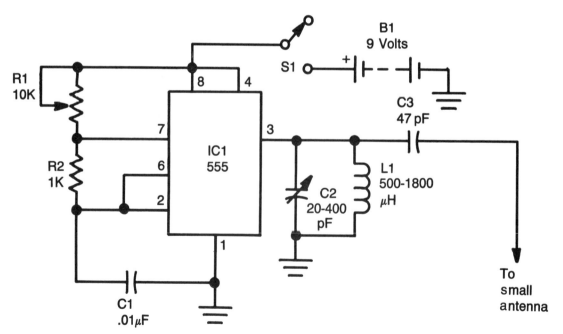

Fig. 2-5.

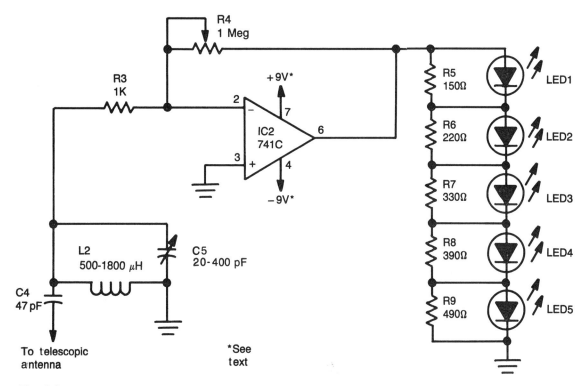

Fig. 2-6.

At the receiver, the incoming signal is tuned by C5 and L2 before being passed on to the 741 IC. That IC amplifies the signal up to 1000 times; the amount of amplification is controlled by adjusting R4, a linear-taper potentiometer (more on that later). The five LED's are used to indicate signal strength, they light up in order (1 to 5) as the signal gets stronger.

The 741 requires two 9-volt batteries for power. The positive terminal of one battery is connected to pin 7. The negative terminal of the other battery is connected to pin 4. The remaining terminals (one positive and one negative) are connected together and grounded.

Keep in mind that you will be carrying the receiver with you as you go about your business, so it should be installed in a case that is as small as possible.

After the devices are built, the receiver and transmitter will need to be tuned. Once that is done, however, you should not need to do it again unless the settings are tampered with. Placing the transmitter and receiver next to each other, detune the receiver so that none of the LED's light. Then tune the transmitter until all of the receiver's LED's light, indicating maximum signal strength. Potentiometer R4 should be adjusted for the minimum amplification that will give you a usable signal. Too much amplification will give you a maximum-strength indication over too wide a range. Separate the receiver and the transmitter (the farther apart they are the better) and adjust R4 until you get a maximum strength reading only when the receiver's antenna is pointed directly at the transmitter. The RF locater is now ready for use.

Since the locater will not be able to work through the metal body of your car, you will need to set up the transmitter so that the signal can radiate through a glassed-in area. That is really not much of a problem. If you are using a hook-up-wire antenna, simply tape the free end to the top of either the front or rear windshield. If you're using a telescopic antenna, place the transmitter on the dashboard and extend the antenna so that it is as long as possible. In either case remember to switch the transmitter on before you leave, and remember that the antenna should be aligned vertically for best results.

To find your car, just extend the telescopic antenna of the receiver to its full length and hold it parallel to the ground. Point the antenna to your far left, then swing it to your far right. Do that until you find in which direction the strongest signal lies, as indicated by the LED's. The antenna will be pointing at your car.

Voltage Freezer

Have you ever wanted to measure the voltage in a tight spot, only to find that before you could read the meter the test probe had slipped and you had to start all over? Having to hold the probe in place and read the meter at the same time is not only inconvenient, but if you slip, you can cause damage.

The circuit described here can solve that problem simply and easily. It reads and stores the voltage, thus freezing the meter reading even after the probes are removed.

The major component of the circuit, shown in Fig. 2-7, is an 8-pin 741C op-amp. The op-amp is configured as a unity-gain voltage follower, with C1 at the input to store the voltage.

The circuit operates as follows: When a voltage is applied across C1, the capacitor charges to that value. When the voltage source is removed, the value is still stored in the capaci-

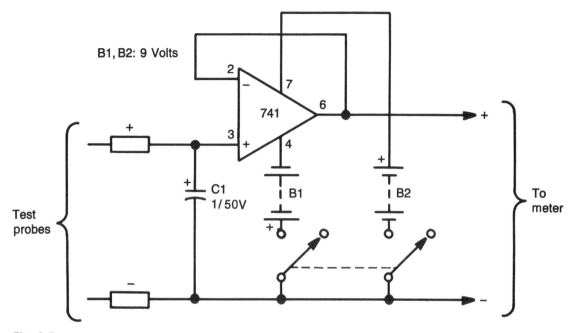

B1, B2: 9 Volts

Test probes

To meter

Fig. 2-7.

tor, and can be read on the meter. While the capacitor *does* discharge, the process takes place very slowly due to the very low loading of the op-amp's high-impedance input. The meter is reset very simply: Just short the probes together; that discharges the capacitor.

Any type of construction can be used for the circuit, since nothing is critical. The only thing you should bear in mind is to use a tantalum capacitor for C1, since it will hold a charge much longer than a relatively leaky aluminum electrolytic. Since no input protection is provided, keep the dc-voltage input below the supply voltage. A DPST switch turns the device on and off, and B1 and B2 are nine-volt transistor batteries. Finally, note that pins 1 and 5 of the 741 op-amp are for offset null; such as, those pins are not required for proper circuit operation and can be ignored.

For better performance, use an LF13741 or a TL081 op-amp in place of the 741—those are JFET devices and offer a much higher input impedance than the 741.

Budget Sound-Effect Generator

Here is a novel use for the Texas Instruments TL507C analog-to-digital (A / D) converter (available from Radio Shack as part No. 276-1789). Although intended for use with 4- and 8-bit microprocessors, this IC provides the "brain" for an incredible sound effects generator. It is relatively easy to build—only a clock and an audio amplifier need to be added. A wide assortment of sounds can be produced—race car, dog's bark, airplane, lion's roar, and more.

The sound effects circuit is shown in Fig. 2-8. A variable clock-pulse generator is made up of two sections of IC1 (a 4069 CMOS hex inverter), R1, S1, and capacitors C1-C6. By adjusting R1, and switching one of the capacitors into the circuit, the clock's pulse rate can be varied over a wide range. That pulse rate can be determined by the formula: Pulse rate = 1 / 1.4RC, where R is in ohms, and C in *farads*. The TL507C (IC2) converts analog signals (in this case the output of IC3, an LM386 audio amplifier) into digital signals at a conversion rate that can be determined from the formula $T = 2N / f$ where

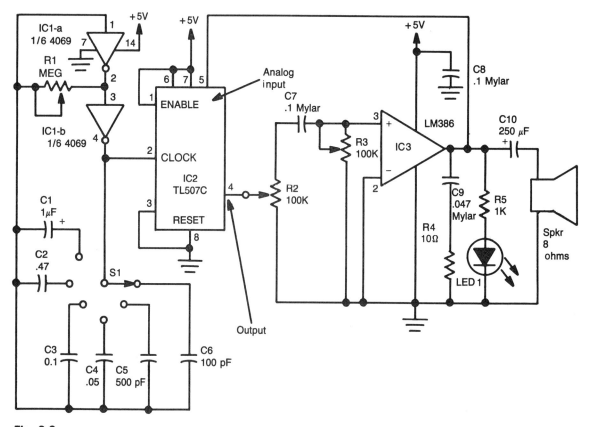

Fig. 2-8.

56

Table 2-1.

Enable	Analog Input	Output
L	X	H
H	$V_1 < 200$ mV	L
H	$V_{RAMP} > V_1 > 200$ mV	H
H	$V_1 > V_{RAMP}$	L
V_1: Analog input to pin 5		
V_{RAMP}: Internal ramp signal		

T is the conversion time, *f* is the clock frequency, and N is the 7-bit output of the binary counter contained in the IC.

The conversion is accomplished using the single-slope method. In short, that involves comparing an internally generated ramp single to the analog input-signal and a 200-mV reference voltage. The application notes for the TL507C show how the relationships between those signals determines the output (see Table 2-1). The RESET pin (pin 8) is held low and the ENABLE pin (pin 1) is held high. That allows continuous conversion operation at a rate determined by the clock frequency and analog input.

The squarewave output from the A / D converter is fed to IC3 through a network consisting of R2, R3, and C7. Resistor R2 controls the amplitude of the pulses. Resistor R3 and capacitor C7 form a variable tone-control filter and a differentiator circuit that converts a squarewave into a spiked waveform. That waveform is amplified by IC3, and the resulting output is fed back into the analog input of IC2 as well as to an eight-ohm speaker.

Indicator LED1 lights to inform you that the power is on, and at the slower clock frequencies it will appear to pulse in time with the sound effects.

By adjusting R1 and selecting one of the six capacitors with S1—thus varying the clock frequency—and by varying R2 and R3, you can produce many sounds.

I built the circuit in a plastic box using perforated construction board and point-to-point wiring, but the method of construction is not critical. I encourage you to experiment with the circuit as I'm sure there are many modifications that could add more fun to it. Those could include LED's coupled to the remaining unused inverters of IC1, additional analog-to-digital converters, different values for R1 and C7, etc.

Crystal Tester

If you frequent hamfests, electronic flea markets, or any other type of surplus outlet, you know the pros and cons of buying from those sources. On the one hand, they're an excellent source of hard-to-get parts as well as a haven for bargain hunters. On the other, however, just about everything is sold "as is," with no guarantee of any kind—it's strictly "let the buyer beware." If you've ever come home with a pile of components, only to find out that half of them were useless, you know that not all bargains are what they seem.

The ideal solution to that problem, of course, is to find some way to weed out the obviously bad parts before you buy them. The circuit I'll be describing here has proved useful for just that purpose when digging through stacks of crystals, as well as in troubleshooting my equipment. It is small, easy-to-build, and will, at a glance, let you know if a particular crystal will oscillate. Let's look at the circuit shown in Fig. 2-9.

Transistor Q1, a 2N3563, and its associated components form an oscillator circuit that will oscillate if, and only if, a good crystal is connected to the test clips. The output from the oscillator is then rectified by the two 1N4148 diodes and filtered

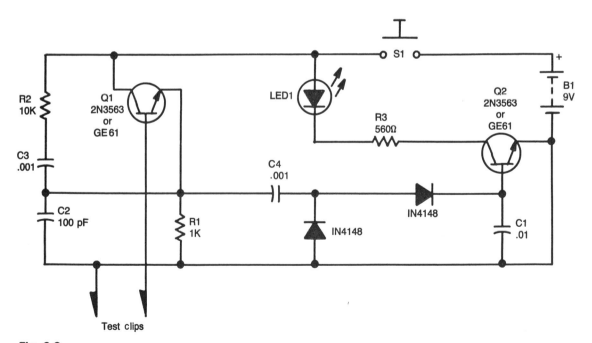

Fig. 2-9.

58

by C1, a .01-μF capacitor. The positive voltage developed across the capacitor is applied to the base of Q2, another 2N3563, causing it to conduct. When that happens, current flows through LED1, causing it to glow. Since only a good crystal will oscillate, a glowing LED indicates that the crystal is indeed OK. The circuit is powered by a standard nine-volt transistor-radio battery and the SPST pushbutton power-switch is included to prolong battery life.

The circuit is easy to build, with size—for easy portability—the only real consideration. While just about any construction technique will work well, it's easiest to use a small piece of perforated construction-board.

To use the crystal tester, simply connect a crystal to the test leads and close the SPST pushbutton power-switch. If the crystal is OK, the LED will glow brightly. If the LED does not glow, or just glows dimly, the crystal is bad and should not be used.

One note on the intended use for the tester is in order here, however. This tester will check any crystal for oscillation. However, it will not necessarily make the crystal oscillate at the frequency that it is supposed to; so you can't use this tester with a frequency counter to test for that. What the circuit *will* do is give you a way to quickly weed out crystals that are obviously bad, and, after all, that is half the battle.

Frequency-Boundary Detector

I'm sure that every electronics experimenter or hobbyist, at one time or another, has needed a device that would indicate whether or not a signal was within a certain frequency range—I know I did when I was working with a switch-mode power supply. I got what I needed by building the frequency-boundary detector whose circuit is shown in Fig. 2-10. (The IC's supply and ground connections are shown in Fig. 2-11.)

The circuit can be used (with LED's or other indicators) to tell you whether or not an input signal is within a certain frequency range. Because you may be hard-pressed to come up with applications for the circuit. I should point out that voltage-to-frequency converters can be used to make the number of applications almost limitless.

The device itself is rather easy to build. It consists of three IC's—a dual monostable multivibrator and two dual D-type flip

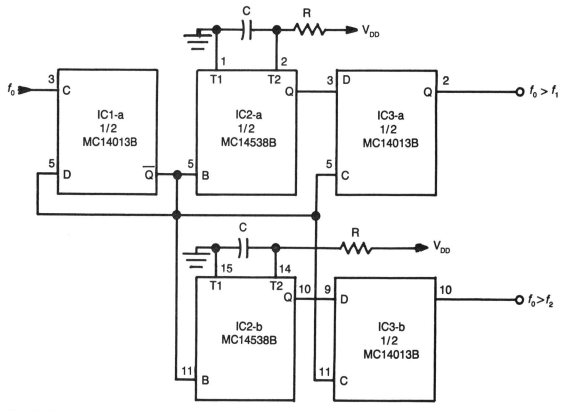

Fig. 2-10.

IC	Pins tied to V_{DD}	Pins tied to ground
IC1	14	4, 6, 7, 8, 9, 19, 11
IC2	3, 13, 16	1,4, 8, 12, 15
IC3	14	4, 6, 7, 8, 10

Fig. 2-11.

flops. The signal whose frequency is in question is fed to the clock input of one of the flip-flops. The $\overline{Q}$ output of that flip-flop (IC1-a) is cross-coupled to its data input so that it acts like a divide-by-two counter. (See the timing diagram in Fig. 2-12.) The trailing edge of the $\overline{Q}$ output is used to trigger the one-shots formed by IC2.

The upper- and lower-frequency boundaries are determined by the two sections of IC2—the dual precision

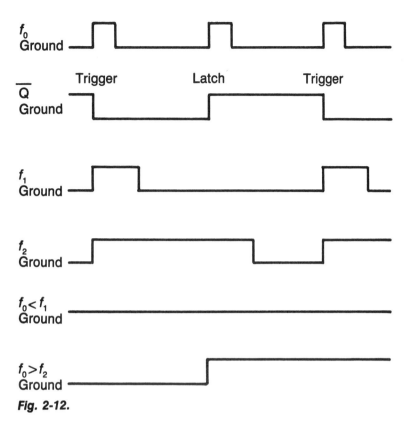

Fig. 2-12.

monostable multivibrator—and their external resistor-capacitor networks. The upper-frequency boundary ($f1$) is set by the output of IC2-a, and the lower-frequency boundary (f 2) is set by the output of IC2-b. The relationship that describes the periods of the outputs of IC2 is: $T = \frac{1}{2}RC$, where T is measured in seconds, R in ohms, and C in farads. However, because IC1-a is used as a divide-by-two-counter, the formula used to determine the period of the upper- and lower-frequency boundaries becomes: $T = RC$.

The frequency of the input to the circuit can be anywhere from dc to 100 kHz. However, you can use the "extra" half of IC1 as another divide-by-two counter and increase the circuit's range to 200 kHz. Then the period of the outputs of IC2 would be represented by: $T = 2RC$.

The states of the outputs of IC2, which determine the upper- and lower-frequency boundaries, are latched by IC3-a and IC3-b respectively. As shown in the timing diagram of Fig. 2-12, the output of IC3-a (which is clocked by the output of

IC1-a) will be high only when the input frequency is less than that of the output of IC2-a (f1). The output of IC3-b will be high only when the frequency of the input is greater than that of the output of IC2-b (f 2). You can use appropriate logic gates to give an "in-bounds" or an "out-of-bounds" indication.

Ultrasonic Pest Repeller

Pest control has been brought into the electronic age by the introduction of the ultrasonic insect repeller. That device is said to repel—not kill—unwanted flying and crawling pests by emitting ultrasonic sound waves that sweep between 65,000 and 25,000 hertz. The sound is apparently rather irritating to them.

I went shopping for one of those "miracle" devices but I was repelled—by their prices, which ranged from $49 to $69. Therefore I decided to design and build my own. The circuit I came up with, which should cost about $20 to build, is shown in Fig. 2-13.

The repeller is designed around a 556 dual timer. One half is operated as an astable multivibrator with an adjustable frequency of 1 to 3 Hz. The second half is also operated as an astable multivibrator but with a fixed free running frequency around 45,000 Hz. The 25–65 kHz sweep is accomplished by coupling the voltage across C2 (the timing capacitor for the first half of the 556) via Q1 to the control voltage terminal (pin 11) of the second half of the 556.

Transistor Q1 serves two purposes: it isolates the timing circuit of the first half of the 556 from pin 11 and it controls an LED indicator. When the first half is operating, timing capacitor C2 continually charges and discharges between ⅓ and ⅔ the supply voltage. Because the base of Q1 is tied to C2, the voltage across C2 will affect the operation of Q1. The voltage at the base of Q1 causes it to conduct, thereby turning on the LED and lowering the control voltage that is applied to pin 11. The lower control voltage causes the output frequency of that half of the timer to increase to around 65 kHz. As C2 is charged toward ⅔ volt, Q1 conducts less and less. That causes the intensity of the LED to decrease and the control voltage applied to pin 11 to increase, because Q1's emitter approaches + V. The increasing control voltage causes the output frequency to decrease from 65 kHz to 25 kHz. That sweep will take from 1 to ⅓ second depending on the setting

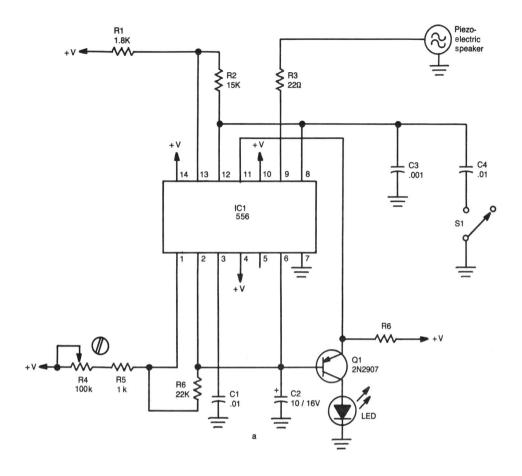

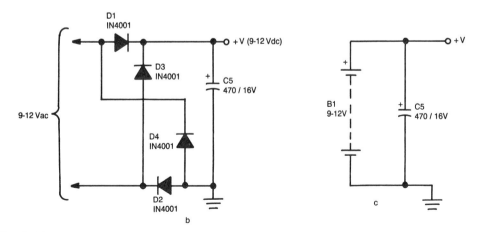

Fig. 2-13.

of R1. Theory has it that periodic adjustment of the sweep rate will prevent the pests from developing an immunity to the sound.

The device that radiates the ultrasonic sound is a piezo tweeter. Radio Shack sells several models ranging in price from $9 to $15.

Because the output of the repeller is above the range of human hearing, it is difficult to determine whether it is operating properly. If S1 is closed, though, the output frequency is lowered so that it can be heard. The output of the piezo tweeter is intense so, if you get tired of the repeller, you can switch C4 permanently into the circuit and turn the repeller into one heck of an alarm.

Proximity Power Switch

How many times have you come home with your arms loaded down with packages and, after great difficulty, managed to turn on the light? And how often have you wished for an easier way? We'll look at a little gadget that allows you to switch an ac device on and off by simply passing a hand (or some other part of the body) near an insulated touchplate (Fig. 2-14).

How It Works

The circuit is powered by a 12.1-volt regulated power supply made up of diodes D3 and D4, capacitor C9, and resistors R12 and R13. The ac voltage is picked directly off the ac line and rectified by D3 (half-wave rectifier). The resulting dc voltage is filtered by C9 and regulated by D4, a 12.1-volt Zener.

Turning to the rest of the circuit we have IC1, which is a 4046 PLL (*Phase-Locked-Loop*). That IC contains a VCO (*Voltage-Controlled-Oscillator*), a source follower, and two phase comparators (which we'll call comparators 1 and 2) with a common input-amplifier. When power is applied to the IC, the VCO outputs a signal at pin 4 that is fed to both comparators via pin 3 for use as a reference. That same signal is also fed to its input at pin 14 through an R-C network consisting of R1, C1, and R2. As long as there's no phase difference detected by the internal phase-comparator between the VCO output and the input signal, the output of IC1 is zero. But when a hand is passed close to the touchplate, body capacitance causes a phase difference. That phase difference is fed to the comparators through the common input amplifier.

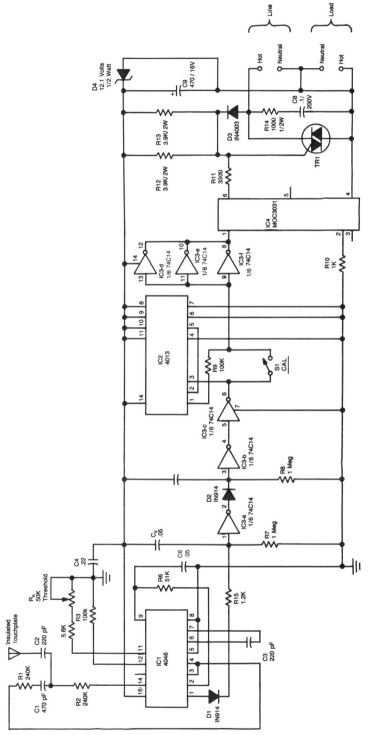

Fig. 2-14.

Comparator 1 now takes that input signal, compares it to the reference and outputs a squarewave signal at pin 2 that is proportional to the difference between the two input signals.

The output of comparator 1 is then filtered and fed to the VCO as an error signal. The error signal causes the VCO to generate an error-correction signal, which is then fed back to the comparators. Comparator 2 then outputs a train or series of pulses that is fed to a pulse-stretching circuit consisting of three of the op-amps contained in IC3, a 74C14 hex Schmitt trigger. (A pulse stretcher is a shaping circuit whose output pulse duration is greater than its input.) The stretcher circuit also provides for debouncing and noise rejection.

After conditioning, the signal is fed to pin 3 of IC2, a 4013 dual D-type flip-flop that's used as a toggle flip-flop for alternate-action switching. By that we mean that the device is turned on or off alternately by the input signal (push on / push off). To accomplish that action, the $\overline{Q}_1$ output of the flip-flop (pin 2) is fed to its DATA input (pin 5). The logic level at the pin 5 is transferred to its Q output at pin 1 during each positive-going transition of the clock pulse.

At this point, the Q output of the flip-flop is fed to the remaining three inverters of the IC3. They are paralleled to provide enough current to drive IC4, an MOC3031 optically-coupled triac driver. When the triac is turned on, current will flow through the triac and the lamp or appliance that is connected to its terminals, turning it on, too.

The touchplate can be made from a one-inch square piece of copper-clad board with a 4-inch piece of wire soldered to it. It should be connected to capacitor C2, as shown. Once you have it all together, the next step is to make sure that it works. To do so, close switch S1 (labeled CAL) and connect a lamp to the load terminals. Now, apply power and adjust potentiometer R5 down from its maximum resistance until the lamp comes on. Bring your hand near the touchplate to insure that the lamp extinguishes. If everything is OK, open S1; the lamp should alternately switch on or off when the touchplate "switch" is activated.

For safer and more stable operation, a low-voltage supply can be used (with R10 adjusted accordingly). However, it will be difficult to mount the switch behind a wall plate if you do that.

The MOC3031 triac driver was selected because of its low drive-current requirements. The less expensive MOC3010 can be used instead of the MOC3031 if zero-voltage switching is not desired. In that case, resistor R11 must be changed to 180 ohms. Switch S1 and resistor R9 are an optional aid and are used in setting up the threshold for operation.

Add-on Scope Multiplexer

Having a dual-trace scope is a luxury that many of us, unfortunately, must do without. However, with the simple circuit we'll describe, you can add dual-trace capability to your single-trace scope at a cost of less than $5. Unfortunately, the device has one major "drawback," it only monitors logic levels (TTL and CMOS); but at that price, who cares!

How It Works

Figure 2-15 shows the multiplexing circuit that lets you view two traces simultaneously. The operation of the unit revolves around three IC's: a 4093 quad NAND, Schmitt-trigger, 4066 quad analog-switch, and a 7555 timer (that is used to gate IC2-b and IC2-c on or off).

The device can be powered from a supply ranging of from 4.5 to 15 volts, and draws less than 2 mA. With a supply of 5 volts, the unit may be used to monitor TTL or CMOS logic-levels. At higher supply voltages (15 volts), it may be used to check only CMOS logic signals.

To make the operation of the unit a little easier to understand, we'll first look at the two input circuits separately and then see how the switching action of the circuit is handled.

When a high is fed to PROBE 1 IN, it is inverted by IC1-a and once again by IC1-b, so that the input to IC2-a is high. That high causes the "switch contacts" in IC2-a to close. With the "contacts" closed, a high-level output is presented to the input of IC2-b.

Meanwhile, let's suppose that a high is fed to PROBE 2 IN. That signal is then inverted by IC1-d and routed to IC2-d, causing its "contacts" to open and the unit to output a logic-level high. The output of IC2-d is then fed to IC2-c.

Unless a gating pulse is presented to both IC2-c and IC2-d, their "contacts" will remain open and no signal will appear at the output. We use the output from pin 3 of IC3 (a 7555 timer) to gate IC2-b and IC2-c. Note that the signal from IC3 is inverted before it is fed to IC2-b but not before it is sent

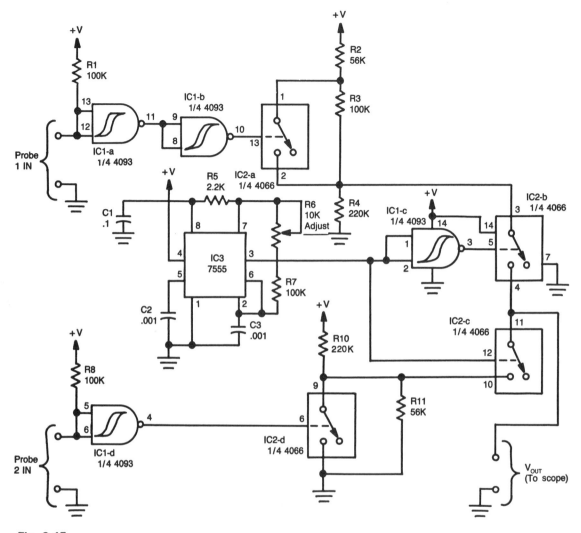

Fig. 2-15.

to IC2-c. Thus, the pulsing output from IC3 will alternately switch the display between probes.

Two voltage-divider networks determine the position that the trace is to be shown on the screen. Because we want to display both signals at the same time, the high and low levels for one probe must be different from the high and low levels for the other probe. For INPUT 1, the divider is made up of resistors R2, R3, and R4, and for the other, R10 and R11.

The addition of R3 in the first voltage-divider circuit increases the voltage level of both the high-level and low-level inputs from probe 1. Thus, the probe-1 signals will be displayed at the top of the signals from probe 2. The probe-1 trace is displayed between the 3- and 4-volt mark, while the probe-2 trace is shown between zero and one-volt. That can be shown by the following formulas, which assume a high level of +5-volts.

Probe 1:

$$\text{High} \cong \frac{R4}{R2 + R4} \text{ (V)}$$

$$= \frac{220K}{56K + 220K} \text{ (5V)} \cong 4V$$

$$\text{Low} \cong \frac{R4}{R2 + R3 + R4} \text{ (V)}$$

$$= \frac{220K}{56K + 100K + 220K} \text{ (5V)} \cong 3V$$

Probe 2:

$$\text{High} \cong \frac{R11}{R10 + R11} \text{ (V)}$$

$$= \frac{56K}{220K + 56K} \text{ (5V)} \cong 1V$$

$$\text{Low} \cong 0V$$

With the scope set to trigger on one input, signals up to 50 kHz can be monitored. That makes the circuit ideal for low to medium speed logic-level inputs. Certain frequencies can cause garbage (harmonics of the sampling frequency) to be displayed; however, adjusting potentiometer R6 will correct that.

Contrast Meter for Photography Buffs

If you're an amateur photographer who enjoys developing his own pictures, you have probably found it necessary to choose the right paper to fill certain requirements. If you're among the fortunate few who can afford a densitometer, then you have nothing to worry about But if you're like most of us and cannot afford that piece of equipment, then perhaps this easy-to-build substitute is for you.

Contrast Meter

Figure 2-16 is a schematic of a contrast meter that can be used to help you choose the right grade of paper for your photographic needs. The circuit, built from readily available parts, will work well with almost any photocell and 1-mA panel meter you choose.

The circuit is powered by a dual 15-volt power supply. If you have trouble in getting the parts to build the power supply, then the design can be modified to use a dual 12-volt sup-

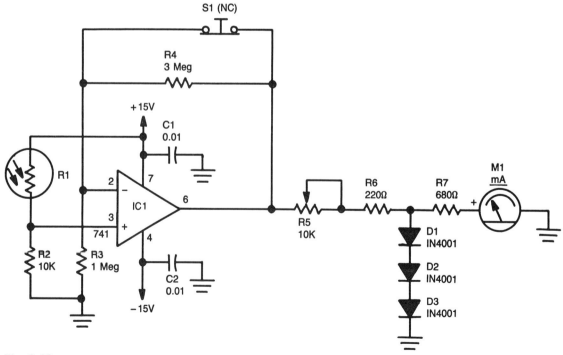

Fig. 2-16.

ply by changing the values of resistors R1, R6, and R7 to 8200 ohms, 180 ohms, and 560 ohms respectively. The only critical components are resistors R3 and R4, which should be tested to ensure a good 1:3 ratio.

How It Works

One leg of the photocell (R1) is tied to the +15-volt supply and the other end is connected to ground through resistor R2, forming a voltage-divider network. The non-inverting input of the 741 op-amp, IC1, is tied to the junction formed by R1 and R2, while its inverting input is grounded through resistor R3. When switch S1 is pressed, another divider network is formed, reducing the voltage applied to the inverting input of the op-amp (more on that later).

When light hits the photocell, its resistance begins to decrease causing a greater voltage drop across R2 and a higher voltage to be presented to the non-inverting input of IC1. That causes IC1 to output a voltage proportional to its two inputs.

The circuit gives a meter reading that depends on the intensity of light hitting photocell R1; therefore, R1 should be mounted in a bottle cap so that the light must pass through a 3/16-inch hole. Potentiometer R5 is used to adjust the circuit for the negative you're working with.

The diode chain, D1-D3, is used to protect the meter in case direct room light hits the photocell. If the dark resistance of the cell is less than about 1 megohm, it may be necessary to use four diodes, instead of the three shown, to get a full scale reading. If the internal resistance of the meter is less than 90 ohms, you may only need two diodes.

The density range of the negative can be expressed as the logarithm (log) of the light intensity (I) through the clearest (shadow) area minus the log of I through the densest (highlight) area. In the form of an equation: density range = $\log (I_s / I_h)$.

By using a simple table of antilogs, you can avoid the need of a log amplifier to determine the correct paper grades corresponding to the specific density ranges.

To use the contrast meter, focus the negative in the enlarger with the lens diaphragm wide open. Then place the photocell under the lightest portion of the negative. Using potentiometer R5, adjust the meter for full-scale deflection. Now, without changing the setting, place the photocell beneath the darkest portion of the negative and read the meter. If the

Table 2-2.

| Density Range | | Paper |
(I$_s$)	(I$_h$)	Grade
>1.4	<4	0
1.2-1.4	4-6	1
1.0-1.2	6-10	2
0.8-1.0	10-16	3
0.6-0.8	16-25	4
<0.6	>25	5

meter now shows less than ten percent of the full-scale reading, it may prove to be very difficult to read accurately.

When that happens, it will be necessary to push switch S1. Pressing S1 removes the short across resistor R4. Because R4 is three times the value of R3, only ¼ the voltage applied to that leg of the circuit will appear at the inverting input of the op-amp, resulting in a reading four times as great. Now simply divide that reading by 4 and compute the ratio of the first and second measurements. And then refer to Table 2-2 to find the right paper grade for that negative.

Making Electronic Music — Automatically

Building electronic circuits to produce musical notes and tones is fast become a favorite pastime of electronics hobbyists. Such circuits have gained popularity among school-aged children for class projects, as well.

Figure 2-17 shows a circuit you can build "from scratch" using any construction method you choose. It's made of several IC's and a few discrete components, and none of the parts used should prove hard to come by.

How It Works

IC1, a 555 timer, is set up as an astable multivibrator to produce the signal that triggers IC2, a 7490 decade counter. That IC, in turn, produces a BCD output that is fed to IC3, a 7445 BCD-to-decimal decoder / driver. Because IC3's output

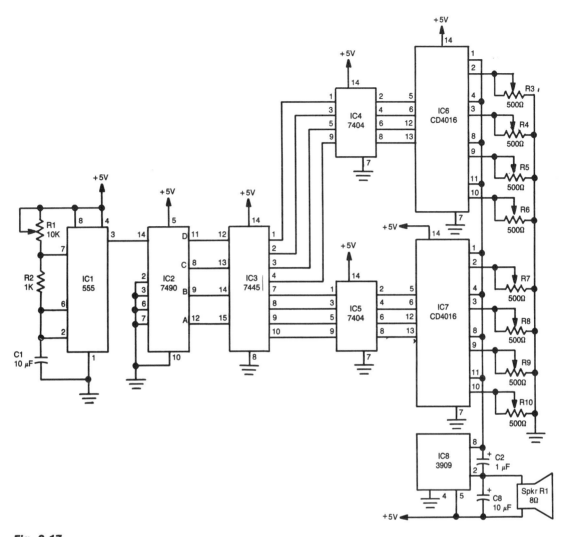

Fig. 2-17.

is positive, it's necessary to invert the signal before feeding it to the rest of the circuit. That's handled by two hex inverters, IC4 and IC5.

Note that the eight outputs of IC3 are divided evenly between the two 7404's. Since there are six inverters in each package, there will be four left over for you to play with. What you do with them is up to you. The outputs of IC4 and IC5 are input to control pins on IC6 and IC7 (DC4016 CMOS quad bilateral switches). As those switches open and close, different resistances (as set by potentiometers R3-R10) are inserted into the sound-generating circuit made up of IC8 and its associated components. Note that IC8, which is 3909 LED flasher also makes a fine sound generator.

The frequency at the outputs of IC6 and IC7 are adjusted to various rates, using potentiometers R3-R10, to produce the desired tones. Capacitors may be placed in series with the potentiometers to produce a sloping sound instead of a straight tone.

Two other tones may be added using the pin 5 and pin 6 outputs of IC3. To do so, simply route those two outputs through inverters and switches as done with the other eight. If you include the extra outputs, it will be necessary to add another CD4016 bilateral switch with potentiometers connected in the same way as the others.

The negative-going output signals of IC6 and IC7 are fed through a common bus to pin 8 of IC8. Filtering for the input signal is provided by capacitor C2. Capacitor C3, connected across the output at pin 2 and the supply controls the speaker output. C3 may be replaced by a potentiometer if desired.

Telephone Off-hook Alarm

Image sitting around waiting for an important telephone call that never comes—and later finding out that the reason that you never received the call was because you (or someone in your household) had left the phone off the hook. Or, how about waiting for a call when, unknown to you, your line has gone dead. Either situation can be frustrating—to say the least.

Well, there's an easy way to solve both of those problems—the *telephone off-hook alarm*. A schematic of the off-hook alarm is shown in Fig. 2-18. That circuit has two indicators to tell you when your telephone receiver—or any extension receiver—is not in its cradle: a lamp, LMP1 and a piezoelectric buzzer, PB1. The lamp lights whenever a receiver is removed from its cradle or the line goes dead.

The buzzer sounds after the phone has been off the hook for a preset period (about 35 minutes), unless the circuit is reset or the receiver is returned to its cradle. The reset switch also allows you to stay on the phone for periods beyond the preset time limit.

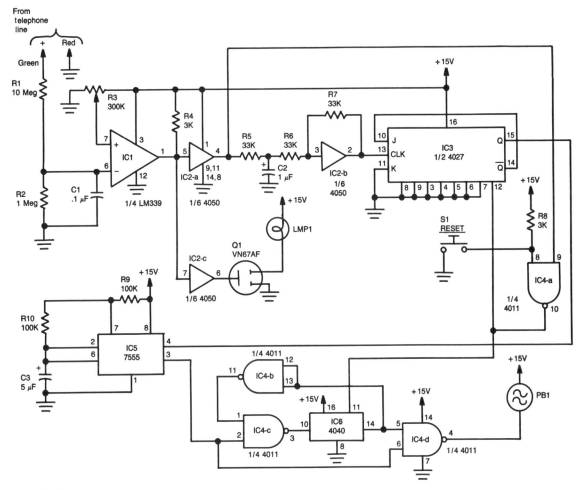

Fig. 2-18.

Circuit Operation

The telephone off-hook alarm is made from several common IC's and a handful of external components. Figure 2-18 shows a voltage divider / lowpass filter at the input to the circuit (made of resistors R1 and R2, and C1). That combination passes a fraction of the telephone line's dc voltage, which is fed to pin 6 of IC1 (an LM339 comparator).

When the phone is on the hook, the telephone line has about 50 volts across it, but once the receiver is lifted from its cradle, that value drops to around 10 volts. Potentiometer R3 is adjusted so that 2 volts is applied to the noninverting

75

input (pin 7) of IC1. Because of the 10-to-1 ratio of R1 and R2, most of the line voltage is dropped across R1.

With the phone on the hook, the voltage appearing at pin 6 is about 4.5 volts, but when the phone is lifted from its cradle that value drops to about 1 volt. When that happens, IC1 outputs a high that follows two paths.

In the first of those paths, that high is fed to IC2-c (⅙ of a 4050 buffer) to provide sufficient drive to turn on (VMOS) transistor Q1. With that transistor turned on, LMP1 lights showing that a receiver is off the hook.

In the other path, the signal is fed to IC2-a, which outputs a high that causes capacitor C2 to charge. When C2 is charged, it triggers IC4-a (¼ of a quad NAND Schmitt trigger) into conduction. The low output of IC4-a toggles JK flip-flop IC3, which then triggers the 7555 (CMOS) timer, IC5. The output of IC4-a is also fed to pin 11, RESET, of IC6 enabling it.

The timer produces a 1-Hz squarewave output that is fed to IC4-d (pin 6), and to IC6 through IC4-c. Now IC6 begins to count, and after about 2048 seconds (or 34 minutes), pin 14 of IC6 goes high. That high is fed to IC4-d, causing its output to go low, which in turn causes the piezoelectric buzzer, PB1, to sound. The circuit is designed so that the buzzer continues sounding until either the phone is hung up, or the circuit is reset by pressing switch S1.

The circuit is powered by a 15-volt, 300-mA supply; and, since the circuit is made of mostly CMOS IC's, power consumption in the standby mode is low. In operation, the lamp draws most of the power. If you want to reduce power consumption even more, the lamp may be replaced with an LED, allowing a power supply with a lower output to be used.

A word of caution: *Do not* earth-ground the circuit. Also, be sure that the power supply is isolated from the ac line.

This Electronic Metronome Emphasizes the Downbeat

Electronic metronomes have long been popular with both electronics experimenters and musicians with a practical bent. All metronomes provide a steady stream of pulses, but few accent the first beat of the measure—the downbeat. The metronome presented here does, and it allows you to vary the counting rate from about 1 to 200 beats per minute. A rotary switch allows you to select an emphasized beat every other beat, every third beat, etc., all the way to once every nine beats.

As shown in Fig. 2-19a, IC1-a and IC1-b form an astable multivibrator whose period of oscillation is approximately equal to 1 / (2.2 × C1 (R1 + R2)). The astable's signal is fed to IC1-c, which buffers the signal for further amplification. The astable's output is also fed to the CLOCK input of IC2, a 4017B (a 4017A is also suitable) decade counter. The IC's Q0 through Q9 outputs go high one at a time for each successive clock pulse received at pin 14. Switch S1 feeds one of those outputs to the 4017B's RESET input; whenever the selected output goes

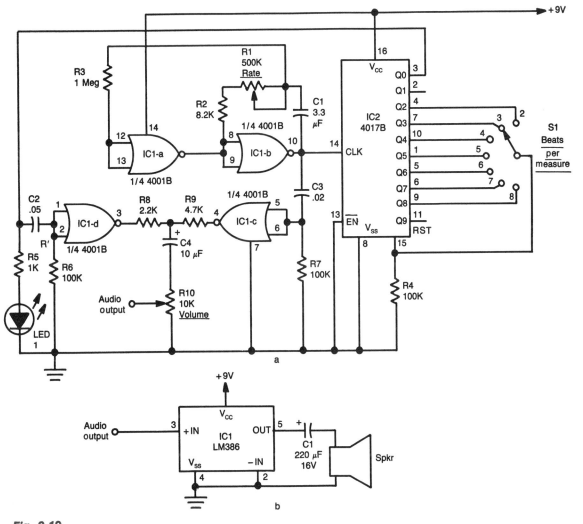

Fig. 2-19.

high, the 4017B restarts its counting cycle. That is what determines the number of beats per measure.

Each time the 4017B is reset, its Q0 output goes high. That signal is fed to LED1 for a visual indication of the start of each measure. The Q0 signal is also fed to IC1-d, another buffer. The signal is also mixed with the astable's free-running output (after buffering by IC1-c). The mixed signal is what provides the extra "oomph" signalling a new measure.

There's not much else to the circuit. The network composed of C2 and R6 sharpens up the downbeat pulse, and the network composed of C3 and R7 sharpens up the free-running pulses. By making C2 larger than C3, the downbeat gets greater emphasis. You may vary the values of those two components to obtain different sound outputs.

The mixed signal is coupled by C4 and R10 to an outside amplifier. You may connect the metronome's output to any convenient amplifier; alternatively, the circuit shown in Fig. 2-19b may be used for that purpose, and will provide a compact, portable metronome.

Construction

The metronome may be built in any convenient manner; just be sure to use sockets for all IC's. After mounting the components, check the board carefully for wiring errors, especially to the battery, a standard nine-volt unit. Use a small speaker, and mount everything in a small plastic case. Just check over your wiring before inserting the IC's in their sockets.

The metronome can be calibrated by marking off ten equally-spaced divisions on the front panel around R1. Attach a knob with a pointer to R1. Turn the metronome on, set R1 to point to the division, and count the number of beats in 20 seconds. Multiply that value by three and record the result. Do the same for each of the other nine divisions, and then transfer the results to the plastic case with rub-on lettering or labels from a labelmaking gun.

Simple Circuit Foils Car Thieves

If you live in a high-crime area and operate an automobile, you're probably afraid that someday your car will be stolen. Well, so was I—until I built and installed the ignition cutoff circuit described here. Since then I sleep much easier. And the circuit has another benefit: Many insurance companies will reduce your rates if you install a cutoff circuit like mine.

What my circuit does is give you manual control over the voltage that goes to your car's coil. If 12 volts doesn't reach the coil, it won't be able to provide the high voltage that fires the spark plugs. The starter will still work, so a would-be car thief will probably think that there's something wrong with your car—and he'll leave it alone.

How It Works

As shown in Fig. 2-20, the heart of the circuit is a 4PDT relay and a hidden pushbutton switch. That switch, the relay's coil, and the ignition switch are all wired in series, so both switches must be closed simultaneously to energize the relay. And since the coil current flows through the relay's contacts, the relay must be energized or current won't get through to the coil.

The relay has four sets of contacts and all four are used. The lowest set of contacts functions as a latch so you don't have to keep your finger on S2 continuously to keep coil

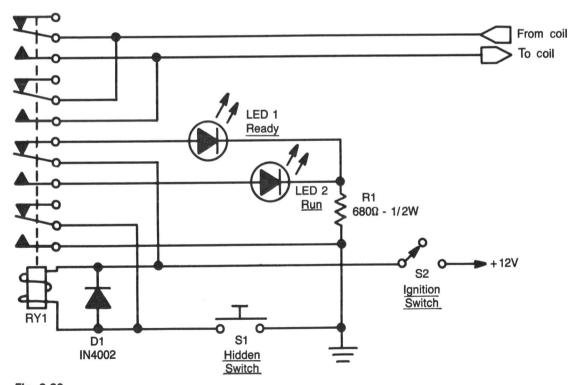

Fig. 2-20.

currently flowing. Those contacts simply keep S2's terminals shorted after the relay has been energized; that short ensures that the relay will remain energized until the ignition switch is opened.

Working upward, the next set of contacts is used to provide visual indication of what's happening. After closing the ignition switch LED1 turns on, but the relay is unenergized, so the car won't start. But when the relay is energized by pressing S2, LED2 lights up to indicate that you should be able to start the car now.

The upper two sets of relay contacts are simply wired in parallel to provide plenty of current-carrying ability for the coil-energizing pulses.

Construction

The circuit can be wired up on a scrap piece of perfboard. None of the parts are critical; just be sure that the relay you use can carry the required coil current. Rather than use two LED's, I used a single two-color device. That way I only had to drill one hole. Be sure to wire a diode across the coil of the relay, and in the orientation shown.

To install the ignition cut-off circuit you'll have to cut several wires in your car's electrical system and splice in leads to your circuit board. You'll have to cut the 12-volt ignition-switch wire and the 12-volt coil-supply wire. When you connect your patches to the circuit board, make sure to use wire that is *at least* as large as the wires you just cut. And make sure to make good connections—electrically and mechanically—to the coil wires.

Sequential Flasher

Here's an easy and inexpensive way to liven up a store window, decorate a Christmas tree, or create a do-nothing toy for the kids. As many as ten lightbulbs can be connected to the circuit and arranged in a circle, or in any other pattern. The lights flash sequentially; when the flash rate is about five or six Hz, an optical illusion of a "running dark spot" is produced.

How It Works

As shown in Fig. 2-21, a 555 timer feeds a 4017 CMOS decade counter. Each of the 4017's first four outputs drives

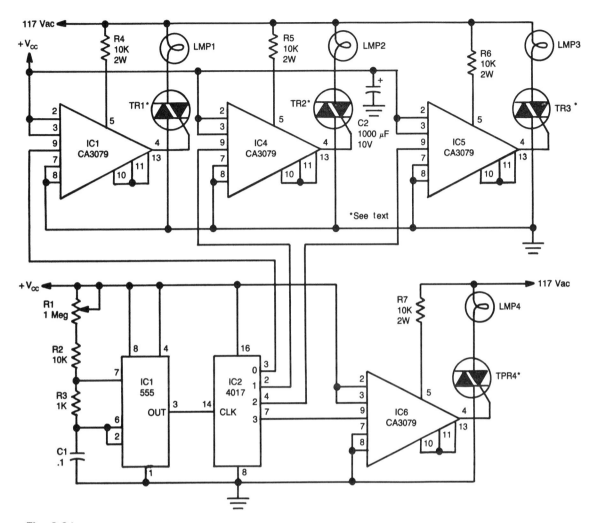

Fig. 2-21.

a CA3079 zero-voltage switch. Pin 9 of the CA3079 is used to inhibit output from pin 4, thereby disabling the string of pulses that IC normally delivers. Those pulses occur every 8.3 ms, i.e., at a rate of 120 Hz. Each pulse has a width of 120 μs.

Due to the action of the CA3079, the lamps connected to the TRI-AC's turn on and off near the zero crossing of the ac waveform. Switching at that point increases lamp life by reducing the inrush of current that would happen if the lamp were turned on near the high point of the ac waveform. In addition, switching at the zero crossing reduces Radio-Frequency Interference (RFI) considerably.

Construction

CAUTION: The CA3079's are driven directly from the 117-volt ac power line, so use care in building the sequential flasher. Keep lead lengths short, use insulated wire and mount the entire circuit in a rigid, insultated enclosure.

We didn't specify part numbers for the TRIAC's, because the type will depend on the lamps you will drive. The TRIAC's will almost certainly require heatsinks; the size of the heatsinks will depend on the amount of power the TRIAC's will have to dissipate, and that depends on the lamps you use.

You'll need a low-voltage source ($+V_{cc}$) to drive the 555, the 4017, and the bias inputs of the CA3079's. One possibility would be to wire up a 7805 regulator circuit and a step-down transformer.

It would also be possible to run the circuit from a 24-volt ac source. Doing so would allow the use of lamps with lower voltage and current ratings. The lower power required by the lamps would also allow use of smaller TRIAC's, smaller heatsinks, and a smaller enclosure. The circuit would also be much safer. See RCA Solid State's *Integrated Circuits for Linear Applications* for more information.

Broadcast-Band RF Amplifier

Unless you own a top-of-the-line receiver or car radio, your AM reception may not be as good as it should be. The reason is that few low- to mid-price receivers and radios include RF amplifiers. By adding one yourself, however, you can improve reception at minimal cost. The RF amp shown here uses readily available parts, has wide bandwidth, and is very stable. In addition, by varying the values of several resistors, you can match the amplifier's input impedance to your antenna, and its output impedance to your radio.

How It Works

The complete schematic is shown in Fig. 2-22. The circuit has a frequency response ranging from 100 Hz to 3 MHZ; gain is about 30 dB.

Field-effect transistors Q1 is configured in the common-source self-biased mode; optional resistor R1 allows you to set the input impedance to any desired value. Commonly, it will be 50 ohms.

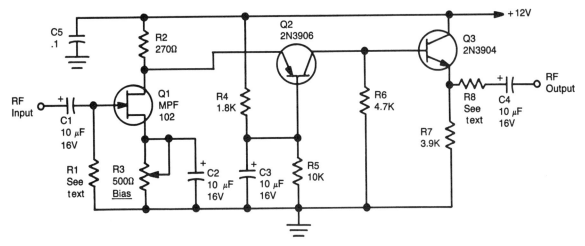

Fig. 2-22.

The signal is then direct-coupled to Q2, a common-base circuit that isolates the input and output stages and provides the amplifier's exceptional stability.

Last, Q3 functions as an emitter-follower, to provide low output impedance (about 50 ohms). If you need higher output impedance, include resistor R8. It will affect impedance according to this formula: $R8 \approx R_{OUT} - 50$. Otherwise, connect output capacitor C4 directly to the emitter of Q3.

Construction

The circuit can be wired up on a piece of perfboard; a PC board is not necessary, although one can be used. However you build the circuit, keep lead lengths short and direct, and separate the input and output stages. You may have space to install the amplifier in your receiver. Otherwise, installing it in a metal case will reduce stray-signal pickup. You'll have to provide appropriate connectors on the case. Connect the amplifier to the antenna and radio using short lengths of coax.

The circuit has only one adjustment. Connect a source of 12-volt dc power to the circuit, and adjust R3 so that there is a 1.6-volt drop across R2.

If you're not sure of the impedance of your antenna, connect a 500-ohm potentiometer for R1, and adjust it for best reception. Then substitute a fixed-value resistor for the potentiometer.

You may want to follow the same procedure with the output circuit (R8), if you're not sure of your receiver's input impedance. Common impedances are 50, 75, and 300 ohms, so the same 500-ohm potentiometer can be used.

You can connect an external antenna through the amplifier to a receiver that has only a ferrite rod antenna. Connect the amplifier's output to a coil composed of 10-15 turns of #30 hookup wire wound around the existing ferrite core, near the existing winding. To obtain best reception, experiment with the number of turns and their placement. You may need to reverse the connections to the coil if output is weak.

SECTION 3

STATE OF
SOLID-STATE

Power Transistor Driver/Amplifier

WHENEVER HI-FI EXPERIMENTERS or audio engineers begin a high-power audio amplifier, they are immediately faced with the question of how to interface the op-amp or low-level discrete voltage-amplifier devices to the power amplifier. The 741 or a similar op-amp—operating from power supplies of ± 6 to ± 12 volts and delivering a maximum of around 5 mA—cannot drive power transistors and Darlington configured devices. In a 50-watt rms amplifier operating from a ± 35-volt supply, those devices require 50 mA or more of drive.

Most approaches to the problem have used two or more discrete driver stages—each with its own power-supply requirements, SOA (*Safe Operating Area*) protection, and short-circuit protection.

Intersil has taken another approach—they have developed a dedicated IC that is the total solution to the problem of driving almost all power transistors with breakdown voltages up to 70 volts. The device—the ICL8063—is a monolithic power-transistor driver and amplifier. It is intended primarily for complementary-symmetry outputs in an audio amplifier and as a driver for linear or rotary actuators, and servo and stepping motors. It is compatible with most op-amps and dedicated devices such as preamps and compandors; taking output levels in the order of ± 11 volts and boosting them to ± 30 volts at 100 mA to drive power transistors. For example, Intersil used 2N3055 (npn) and 2N3791 (pnp) as the output transistors in their data-sheet circuits. The ICL8063 includes built-in ± 13-volt regulated outputs to power op-amps or other external devices. Therefore, only ± 30-volt supplies are needed for a complete power amplifier.

Using the ICL3068, we can build a power amplifier delivering ± 2 amps at ± 25 volts with only three additional discrete devices (a pre-driver and two power transistors) and as few as eight passive components. The slew rate of the power amplifier is the same as that of the 741 pre-driver by itself; except that the output current can slew up to 2 amps at 1V / μs. Other factors such a common-mode rejection ratio (CMRR), input current, voltage offset and power-supply rejection ratio (PSRR) are also the same as for the 741 op-amp. Typically three 1000-pF (.001 μF) compensating capacitors are used to insure good stability down to unity gain. The circuit will drive a 1000-pF load (as might be represented by 30 feet of RG-58 coaxial cable) in line-drive application, without problems. Quiescent current is only 30 mA from a ±30-volt power supply.

A $20-Per-Channel 50-Watt Amplifier

Figure 3-1 is the schematic of a power amplifier using the ICL8063 to drive 2N3055 and 2N3791 power transistors to 50 watts into an 8-ohm load. (The pinout of the ICL8063 is shown in Fig. 3-2.) The first 741 is a preamp for FM tuner and phonograph inputs. The phono input has RIAA (*Recording Industry Association of America*) equalization. The second 741 is a pre-driver for the ICL8063. The complementary-symmetry-output transistor stage delivers 56 volts P-P (50 watts rms) into an 8-ohm speaker. Distortion is less than 0.1% up to about 100 Hz and increases to about 1% at 20 kHz.

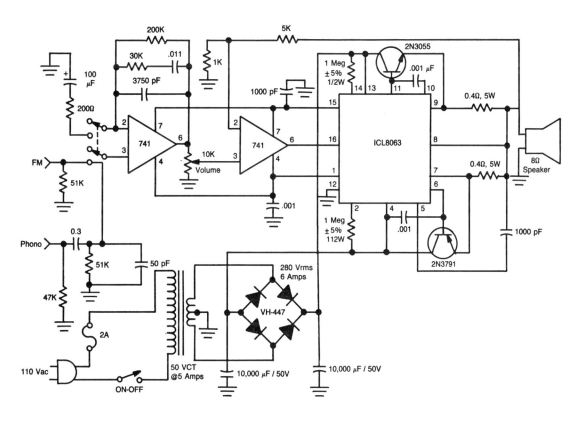

Fig. 3-1.

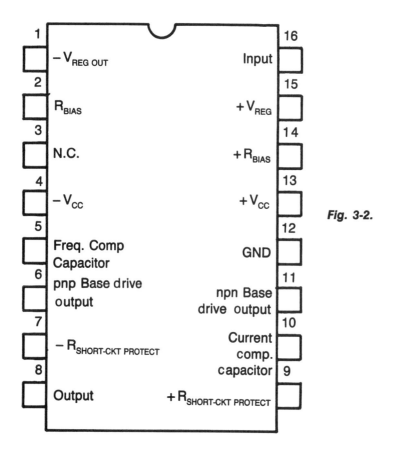

1 $-V_{REG\ OUT}$		16 Input
2 R_{BIAS}		15 $+V_{REG}$
3 N.C.		14 $+R_{BIAS}$
4 $-V_{CC}$		13 $+V_{CC}$
5 Freq. Comp Capacitor		12 GND
6 pnp Base drive output		11 npn Base drive output
7 $-R_{SHORT\text{-}CKT\ PROTECT}$		10 Current comp. capacitor
8 Output		9 $+R_{SHORT\text{-}CKT\ PROTECT}$

Fig. 3-2.

The 0.4-ohm resistors limit the maximum output current that can be drawn. The 1-megohm biasing resistors (between pins 2 and 4 and 13 and 14) are based on $V_{cc} = \pm 30$ volts and guarantee adequate performance when driving dc motors, programmable power supplies, and power DAC's. You can decrease V_{cc} from ± 30 to ± 5 volts in 5-volt steps by using 1 megohm, 680k, 500k, 300k, 150k, and 62k biasing resistors.

When selecting the output transistors for the amplifier, make sure that their beta (hfe) does not exceed 150 at $I_c = 20$ mA and $V_{CE} = 30$ mV. The output terminal can be shorted to ground for an indefinite period as long as the transistors have adequate heat sinks.

Dual High-Speed Rectifiers

The RUR-D1610, -D1615, -D1620 series is a new family of RCA ultra-high-speed dual-chip rectifiers intended as output rectifiers and fly-wheel diodes in high-frequency pulse-width-mounted power supplies and switching regulators. The devices feature a current-carrying capacity of 16 amps per diode and a recovery time of less than 35 ns. Maximum forward voltage drop (at 25°C and full rated current) is 0.95V.

The low stored-charge and fast recovery of the RUR series minimizes electrical noise and, in many circuits, reduces the turn-on power dissipation of associated switching transistors. Breakdown voltages for the RUR-D1610, -D1615, and -D1620 are 100, 150, and 200 volts, respectively. The devices are in steel TO-204M packages.

New Voltage Regulators

Motorola has introduced a series of three-terminal negative-voltage regulators capable of supplying in excess of 1.5 amps over an output-voltage range adjustable from −1.2 to −37 volts. These voltage regulators—the LM137 / 237 / 337—are easy to use and require only two external resistors to set the desired output voltage. Added features found in the new series include internal current limiting, thermal shutdown, and safe-area compensation.

The LM337T, 337H, and 337K are packaged in TO-220, TO-39 and TO3 housings, respectively. Their temperature range is 0°C to C +125°C. The LM237 and LM237K are in TO-39 and TO-3 packages, respectively and are developed to operate in the −25°C to +125°C range. The LM137H and LM137K operate over a −55°C to +150°C temperature range and are in TO-39 and TO-3 packages, respectively.

Memory Design Kit

The TMS4500-Kit from Texas Instruments is the firm's *MOS Memory Design Kit* that comes with enough devices and supporting literature to permit you to design a 32K byte memory system. The kit contains four TMS4416's—the firm's DRAM (*Dynamic Random Access Memory*)—and the TMS4500A DRAM controller featuring an address multiplexer, refresh counter, timing and control circuits, and logic for microprocessor access and memory-refresh sequences.

Power Op Amp IC's

Power opamps are relatively new arrivals on the semiconductor scene, and they've brought with them several interesting applications. One such device is the L272 dual, power op amp from SGS. A pinout of that device is shown in Fig. 3-3a, and a schematic of one op amp contained in the package is shown in Fig. 3-3b.

Housed in a 16-pin, power-DIP package, the IC is intended for various applications including servo amplifiers and power supplies. The L272 can operate from either single or split power-supplies ranging from 4- to 28-volts dc. Its output current can range up to 1 ampere. Other pertinent electrical characteristics are given in the table in Fig. 3-4.

The wide voltage range of the L272, along with its current-handling capabilities, make the unit ideal for controlling low-voltage dc motors. Therefore, it should find many uses in the fields of remote control and robotics. The data sheet includes such applications as a motor current control and a bidirectional dc-motor control (with or without microprocessor-compatible inputs).

Figure 3-5 shows a circuit using the L272 that's designed as a position control for automobile headlights. However, it may be used along with a surplus gear-motor for positioning heavy ham or CB beam antennas. And if you're interested in building satellite TVRO equipment, the circuit might be used in a motor-drive system to aim the antenna dish at various satellites.

The circuit is a bridge arrangement with two voltage-divider networks. The noninverting input (pin 7) of IC1-a is connected to one divider network through switch S1 (the position-selector switch). The noninverting input of IC1-b at pin 6 is connected to a second resistor string through switch S2 (which is driven by the positioning motor).

When identical voltages are applied to the two noninverting inputs of the op-amps, the bridge is balanced and the motor is at rest. When the bridge balance is upset (by changing the position of S1), the motor turns in the direction that brings the bridge back into balance and moves the controlled device into the position selected by S1.

Four-Channel Analog Switches

Dual four-channel analog switches, the LM1037 and 1038, for source selection in stereo-audio equipment and for use in a wide range of industrial, automotive, multiplexing, and sampling applications have been announced by National Semiconductor Corp.

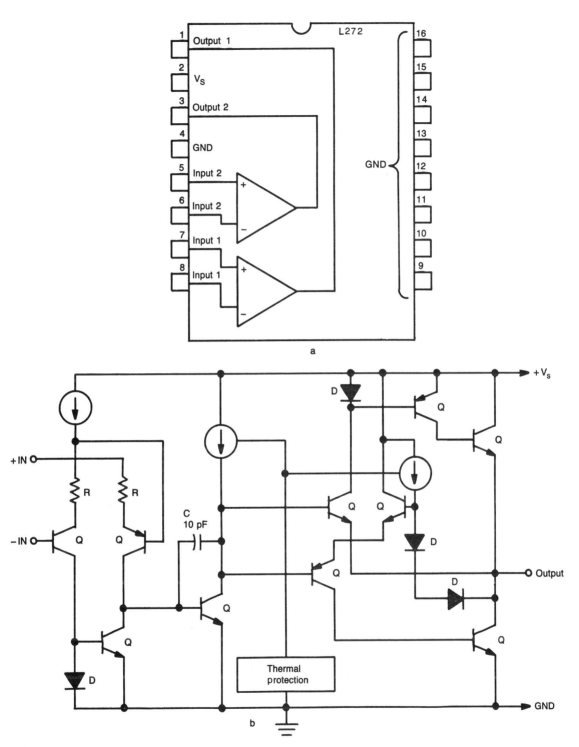

Fig. 3-3.

Parameter		Test Conditions		Min.	Typ.	Max.	Unit
V_S	Supply Voltage			4		28	V
I_d	Quiescent Drain Current				5.5	12	mA
I_b	Input Bias Current				0.5	25	μA
V_{OS}	Input Offset Voltage				15		mV
I_{OS}	Input Offset Current				50	250	nA
SR	Slew-Rate	$G_V = 1$			1		V / μs
B	Gain-Bandwidth Product				350		kHz
V_0	Output Voltage Swing	$f = 1$ kHz	$I_P = 0.1$A		23		V P-P
			$I_P = 0.5$A		22.5		
R_1	Input Resistance			500,000			Ω
G_V	Voltage Gain (Open Loop)				70		db
S_N	Input Noise Voltage	B = 10 to 10,000 Hz			5		μV
I_N	Input Noise Current	B = 10 to 10,000 Hz			200		pA
CMR	Common Mode Rejection				70		dB
SVR	Supply Voltage Rejection	$f_{RIPPLE} = 100$ Hz	Single Supply		70		dB
			Split Supply		62		dB
T_{SD}	Thermal Shutdown Junction Temperature				160		°C

*$V_S = 24$V, $T_{AMB} = 25$°C Unless Otherwise Specified

Fig. 3-4.

The LM1037 units have four control inputs to select any one of four possible stereo-input signals. All channels are muted internally when no input is selected. Electronic controls simplify the routing of audio signals, and allow dc selection with low noise and low distortion. The high-input and low-output impedances make the LM1037 advantageous when compared to CMOS types.

The LM1038 is similar to the LM1037, except that it is designed to be controlled by a microprocessor. Available in 18-pin plastic DIP's.

IC Temperature Sensors

Temperature sensors have come a long way since the invention of the mercury-bulb thermometer. In the past, electronic devices that indicated changes in temperature were often based on the principle that a device's resistance varies as temperature changes. That variable resistance would cause a change in voltage that could be sampled, scaled, and output in human-readable form. Thermocouples, for example, work

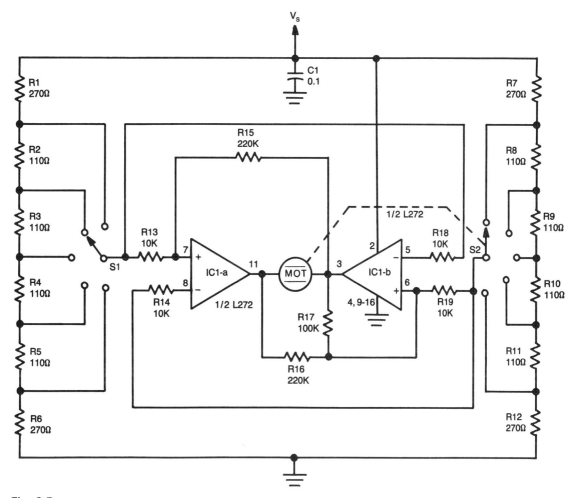

Fig. 3-5.

according to such principles. But the latest in electronic temperature sensing is based on integrated-circuit technology.

National Semiconductor has introduced two series of precision IC temperature sensors; each series consists of five different devices with different temperature ranges. In each device, output voltage is linearly proportional to temperature. The LM35 series is used for Celsius readings, and the LM34 series for Fahrenheit readings.

All devices in the series are trimmed and calibrated during manufacturing to provide high accuracy and linearity; hence the circuit designer need not provide either calibration or trimming. The LM35 series features accuracy of ± 0.25-degree at room temperature, and ± 0.75-degree over the full -55 to

+150°C range. The accuracy of the Fahrenheit units is ±0.5-degree at room temperature, and ±1.5-degree over the −50°F to +300°F range. Those devices have one advantage over similar units calibrated in degrees Kelvin. It is unnecessary to subtract or to null out a large constant in order to obtain readings directly in either degrees Celsius or degrees Fahrenheit.

All devices feature low output impedance, linear output, and precise inherent calibration, all of which make interfacing to readouts or control circuits especially easy. They can be powered by single-ended supplies, ranging from +4 to +30 volts, or by split (plus and minus) supplies.

Quiescent current drain is very low. The LM35 series draws 56 μA from a 5- to 30-volt supply at 25°C. The LM34 series draws 75 μA. Self-heating due to thermal resistance is less than 0.1°C or 0.2°F in still air. Devices in the series measure temperatures ranging all the way from −55°C to +150°C (LM35, LM35A), and −50°F to +300°F (LM34, LM34A). Other models are available with more restricted ranges.

Thermal resistance of the LM35 in the TO-46 package is 400°C / W junction to ambient and 24° C / W junction to case. Thermal resistance in the TO-92 package is 180°C / W junction to ambient. The LM34 has a thermal resistance of 292°F / W junction to ambient, and 43°F / W junction to case in a TO-46 metal-can package. Thermal resistance is 324°F / W junction to ambient in the TO-92 plastic package.

All devices in the series produce a linear 10.0 mV / degree (°C or °F) output over the range of +2°C to +150°C for the LM35, and +5°F to +300°F for the LM34. Figure 3-6 shows the LM35 as an expanded-scale Fahrenheit thermometer with a +50°F to +80°F range.

Figure 3-7 shows the LM34 used as a bar-graph temperature display that displays temperatures ranging from +67 to +86°F. Two LM3914 bar-display LED drivers control twenty LED's. All fixed resistors are 1% or 2% film types. Adjust trimmer resistor R11 so that the voltage at pin 8 of IC3 is 3.525 volts; adjust R8 so that the voltage at pin 4 of IC2 is 2.725 volts; and adjust R5 so that the voltage at the output of IC1 is 0.085 volts + 40mV / °F × T_A (ambient temperature). For example, for an ambient temperature of +80°F, V = 0.085 + (0.04 × 80) = 0.085 + 3.2 = 3.285 volts.

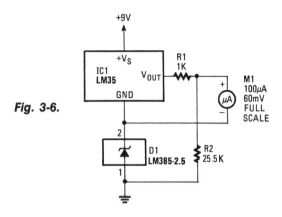

Fig. 3-6.

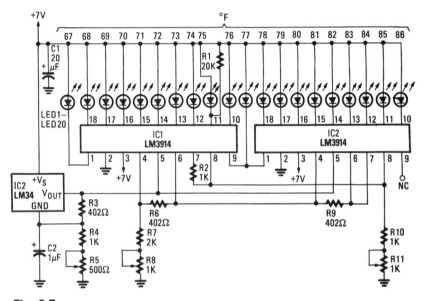

Fig. 3-7.

The data sheets on these two device families come with complete specifications and a dozen or so practical circuit applications.

New Transient Suppressor

The *Surgector* is a new type of device capable of diverting dangerous transient energy away from sensitive electronic equipment like telephones, computers, and other types of equipment subject to sudden voltage surges.

The monolithic device is a thyristor whose gate region contains a special diffused section that functions as a Zener diode, and that also permits anode-voltage turn-on of the device. It is claimed that this feature provides high-speed protection not available with many transient-protection devices presently used.

Risetimes of transient voltage spikes are often very fast; for example, lightning often produces transients with risetimes exceeding 1000 volts per microsecond. Gas tubes and many other protective devices cannot act fast enough to limit the voltage across the protected circuits. The *Surgector* uses Zener action to clamp the voltage until the integral thyristor turns on and drops the voltage to a safe value. In most cases the peak voltage reaching the protected circuitry does not exceed 130% of normal operating voltage. For example, it is claimed that a lightning surge with a risetime of 1500 volts / μs is clipped at about 100 volts.

Presently, RCA offers four types of Surgector: the SGTO3U13, SGTO6U13, SGT23U13 (2-terminal devices), and the SGT10S10 (a 3-terminal device). The 2-terminal devices are available with voltage ratings of 30, 58, and 226 volts. Those ratings refer to the voltage that can be continuously applied without tripping the device.

When a high-voltage transient arrives, the Zener diode in the gate region of the SCR conducts, and that turns the SCR on. The transient is thereby clamped to the forward drop of the SCR so the protected circuitry cannot be damaged. After the transient has passed, and after normal circuit current has dropped below the holding current of the Surgector, the device turns off, and normal circuit operation resumes. The devices have holding currents above 100 mA, and that insures they will operate in average telecommunication circuits.

The SGT10S10 is unidirectional, and its third terminal allows the user to control the SCR's turn-on voltage. That voltage is normally 100 volts, but, by using external gate-control circuitry, voltages less than 100 can be used to trigger the device.

All devices in the SGT series are housed in modified TO-202 plastic packages, whose small size makes them ideal for telephone handsets and PBX's. However, the SGT devices' low cost and high speed also make them suitable for applications in computers, alarm systems, TV, aircraft electronics, and CATV.

One-IC AM Receiver

The ZN416E is the latest-addition to Ferranti's line of single-IC AM broadcast-band receivers. Similar to the ZN415E in packaging and pin-out, the new device is a buffered-output version of the TO-92 style ZN414Z. A typical ZN416E delivers 120-mV rms into a 64-ohm load.

Powered by a single 1.5-volt dry cell, the device may be used in a wide range of applications, including personal receivers, novelty radios, remote telephones, and radio-control circuits. The ZN416E, like others in its family, can be used as the IF-strip and detector of an AM superheterodyne receiver.

The ZN416E features a 150-kHz to 3.0-MHz input-frequency range, and it includes an RF amplifier, a detector, AGC and an audio amplifier. The output stages provide 18-dB voltage gain that is suitable for direct-drive headphone applications.

Overvoltage Protection

High voltages strike fear in the heart of circuit designers because damage by a high-voltage condition can easily cause erratic circuit operation or even catastrophic component failure. To protect circuits from overvoltage conditions, Motorola has introduced four IC's that work with both positive and negative supplies. Those IC's sense the overvoltage condition and almost instantly "crowbar" (short circuit) the power-supply line; the dangerous voltage is thereby reduced before sensitive circuitry can be damaged. One nice feature of the new IC's is that an external capacitor may be used to program a delay between the onset of the overvoltage condition and the tripping of the crowbar. That delay provides noise immunity. The IC's also have circuitry that eliminates trip-voltage and temperature-drift errors due to SCR gate variations.

The MC34061 is a three-terminal device in a TO-92 plastic package. The basic MC34061 offers a ±2% trip-voltage tolerance. The corresponding figure for the MC34061 / A is ±1%, and its other key parameters have been tightened. Other features of the MC34061 include:

- 200-mA SCR gate drive
- 2.5-volt sense voltage
- 250-mV hysteresis
- 4-41 volt supply voltage

A block diagram of the MC34061 is shown, along with a typical application, in Fig. 3-8a. The voltage at the comparator's inverting input (pin 3) is $(V_{CC} \times R2) / (R1 + R2)$, while the voltage at the noninverting input is $V_{CC} - 2.5$ volts. Therefore, the voltage divider (R1 and R2) sets the sense-trip level, and the comparator's output is a function of V_{CC}.

The trip voltage (V_{TRIP}) equals $2.5(R1 + R2) / R1$. When V_{CC} is less than V_{TRIP}, the output transistor is *off*. When V_{CC} is greater, the output transistor is *on*.

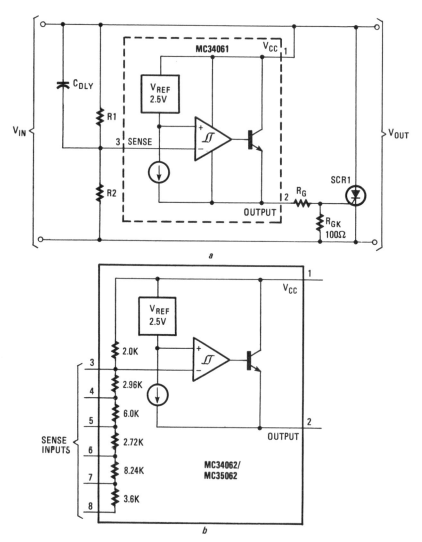

Fig. 3-8.

In the *off* state, a small current (the sum of the reference- and comparator-supply currents) is available at pin 2. Resistor R_{GK} may be used to shunt that current away from the driven circuit. A value of 100 ohms reduces the off-state drive to about 60 mV.

In the *on* state, the device becomes a current source capable of saturating to within 2.0 volts of V_{CC}. So, if the device must drive a high-impedance load, you'll have to clamp the output to at least 3.0 volts below V_{CC}.

Resistor R_G should be connected in series with the SCR's gate when you use a power supply greater than eleven volts. That gate resistor limits the power dissipated in the IC to about two watts. It also protects the IC if the SCR fails. The data sheets for the MC34061 supply detailed information, including nomographs, on selecting an appropriate SCR and gate resistor.

The delay provided by capacitor C_{DLY} is a function of R1, R2, C_{DLY}, as well as the nominal supply voltage and the value of the overvoltage. The magnitude of the overvoltage condition determines the rate at which C_{DLY} charges up to the reference voltage (2.5 volts). So, for given values of R1, R2, and C_{DLY}, the delay decreases as overvoltage increases. The time (in milliseconds) may be found from this equation:

$$T_{DLY} = \frac{R1 \times R2 \times C_{DLY}}{R1 + R2} \times$$

$$\ln \left(\frac{V_{CC} - V_{CC\,(NOM)}}{V_{CC} - V_{TRIP}} \right)$$

Motorola provides a nomograph that simplifies determining the time delay for various values of C_{DLY} at supply voltages ranging from 6.3 to 40 volts. In a typical 5-volt supply, R1 = 1.8K, R2 = 2.7K, V_{CC} = 5.0 volts, and V_{TRIP} = 6.25 volts.

The MC34062 and MC35062 are similar devices with built-in trip-point sensing. They come in eight-pin dual-in-line packages (DIP's). The MC35062U comes in a ceramic DIP and operates over the military temperature range of -55 to $+125°C$. The MC34062P1 (ceramic) and MC34062U (plastic) operate over the commercial temperature range of 0 to $+70°C$.

The MC34062 and MC35062 are very similar to the MC34061. They differ from it in that they include a built-in voltage divider, as shown in Fig. 3-8b, that allows the user to program a trip voltage ranging from 3.5- to 40-volts dc. By connecting the input voltage to a single pin, an MC34062 / MC35062 can trip at 5, 10, 12, 15, 24, or 28 volts. By inter-connecting pins, grounding them, or both, the user can select 120 other trip voltages ranging from 3.483 to 39.064 volts.

Single-chip Sync/Sweep Circuit

A complete television sync / sweep generator in a single IC package has been developed by RCA's Solid State division. The CA3218E Vertical-Countdown Digital-Sync System is a significant advance over the sync and sweep circuits that have been recognized as industry standards for more than a decade. The IC is designed for operation in 525-line televisions, monitors, and video display equipment. The IC's unique vertical-countdown design uses a 10-stage counter and logic circuits to improve noise immunity and to permit elimination of the vertical-hold control.

The CA3218E also provides composite blanking and burst-gate output signals that can be summed in a simple external RC network to produce the "sandcastle" waveform signal needed for the operation of chroma and luminance circuits in color-TV receivers.

The device works with both standard and non-standard sync signals. An automatic mode-recognition circuit forces the IC to operate in the non-synchronous mode when the incoming composite video signal has a scanning rate other than the standard 60-Hz, 525-line format. The CA3218E might be used in a circuit like the one shown in Fig. 3-9.

How It Works

The IC's internal master oscillator is controlled by an external RC network that is connected to pin 5. The oscillator runs at eight times the horizontal rate. The signal is divided several times and fed to other portions of the IC. For example, a divide-by-8 output goes to the horizontal amplifier output (pin 8). A phase-locking AFC circuit controls the precise frequency

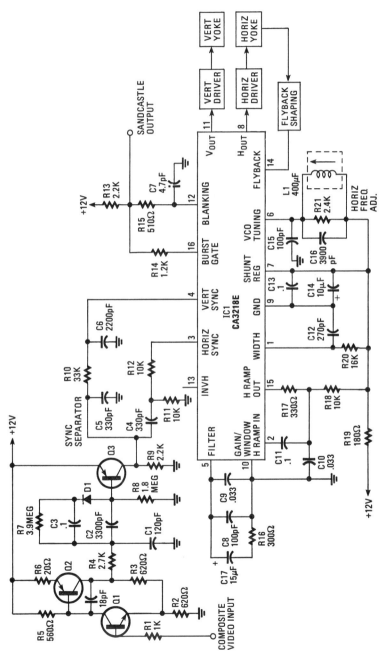

Fig. 3-9.

101

of the master oscillator. The horizontal ramp signal input (pin 2) is derived from the flyback pulse that appears at pin 15. Then it's fed to a phase detector where it is compared to the horizontal sync fed in at pin 3. The phase detector generates a correction voltage that keeps the master oscillator phase-locked to the correct frequency.

Divide-by-2 and divide-by-4 outputs of the horizontal divider drive the 10-stage vertical counter. The digital countdown system and associated logic circuits provide good noise immunity and eliminate the need for a vertical-hold control.

The GAIN / WINDOW input (pin 10) is a logic input that controls the vertical-sync "window" during which the system looks for the occurrence of a vertical-sync pulse on the incoming signal.

Upon receipt of the 464th or the 512th clock pulse (according to whether pin 10 is low or high, respectively), the vertical-sync window is "opened." The end of the sync window is marked by the arrival of the 568th or the 592nd pulse, again according to whether pin 10 is high or low.

If the incoming vertical-sync pulse coincides regularly with the 525th clock pulse, vertical blanking and sweep signals are generated in the standard sync mode. If the incoming vertical sync does not coincide with the 525th clock pulse or if it is masked by noise, the 10-stage counter continues to supply an output at the 525th clock pulse. A three-bit counter counts the number of consecutive fields during which a sync pulse does not arrive simultaneously with the 525th clock pulse. If a sync pulse does not coincide with the 525th clock pulse for eight consecutive fields, the 3-bit counter activates a circuit that switches operation to non-standard sync.

When the circuit operates in the non-standard sync mode, the incoming vertical pulse initiates vertical sweep when the input pulse coincides with any clock pulse within the sync window except the 525th. If no vertical-sync pulse arrives during the window time, the system freewheels at a frequency determined by the 568th or 592nd clock pulse, depending on the state of pin 10.

The CA3218E generates a composite blanking signal and a burst-gate keying pulse at pins 12 and 16, respectively. Those signals can be summed in a simple external resistive network (R13, R14, and R15 in Fig. 3-9) to produce the "sandcastle" signal that is used to drive chroma / luminance IC's like RCA's CA3220E. The CA3218E also generates an inverted horizontal sync signal (at pin 13) that can be used to drive the CA3224E Automatic Picture Tube Bias IC.

"Chop-Amps"

When you need an op-amp that has low input-bias current, extremely low input-offset voltage, excellent long-term stability, and a good temperature coefficient, two devices to be considered are the LTC1052 and the LTC7652 chopper-stabilized op-amps from Linear Technology Corporation. The chopper-stabilization scheme constantly monitors and corrects offset-voltage errors that develop over time, along with errors due to variations in temperature and common-mode voltage. That error correction, coupled with input currents in the picoampere range, gives those devices exemplary performance and makes them worthy of use in many applications of consideration.

The LTC1052 and the LTC7652, sometimes called "discrete-time" or "sampled-data" amplifiers, are available in the metal can "H" package, and in 8- and 14-pin hermetic and plastic DIP's. In various packaging configurations, the op-amps are direct replacements for similar devices from Intersil.

Both op-amps feature a maximum offset voltage of five μV and a maximum offset drift of 0.05 μV / °C. In addition, maximum input bias current is 30 pA. Other specifications include: minimum gain of 120 dB; minimum CMRR of 120 dB; and minimum PSRR of 120 dB. Either op-amp can be operated from a single-ended power supply (maximum of 18 volts); and both are rated to operate over the -40 to $+85$°C range.

Chopper stabilization works like this. As shown in Fig. 3-10, the main amplifier is connected at all times between the input and the output of a circuit. A nulling amplifier, controlled by a chopping-frequency oscillator and a clock circuit, is cross-connected to the main amplifier circuit—input to output and output to input. The chopper can be thought of as several ganged switches (S1-S3), and the amplifiers as trans-conductance amplifiers. Two external capacitors (C1-C2) function as sample-and-hold circuits; the value of each capacitor normally ranges from 0.1 to 1.0 μF.

The null amp alternately nulls itself and the main amplifier. During half of the clock cycle (the auto-zero phase), the null amplifier inputs are shorted and a voltage equal to the

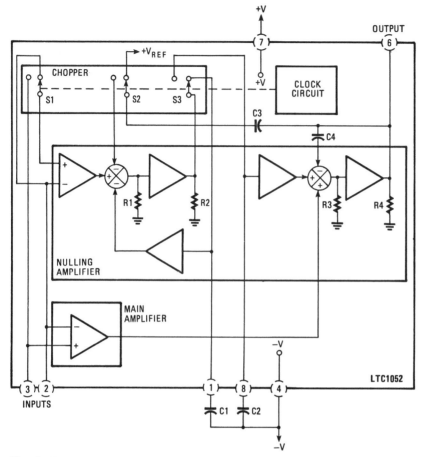

Fig. 3-10.

amplifier's offset voltage is stored on capacitor C1. During the following half-cycle of the clock (the sampling phase), the main amplifier's inverting and non-inverting inputs are compared and a voltage equal to the input-offset voltage is stored on capacitor C2. In other words, the two capacitors store the potentials needed to null the offset voltages of both the null and the main amplifiers. The nulling scheme operates over the full common-mode and supply-voltage ranges, so CMRR, PSRR, and large-signal voltage gain are extremely high. A constant dc voltage for nulling the input offset is produced by switching between the sampling and the nulling phases at a frequency much lower than the frequency of the input signal.

When the null amplifier inputs are shorted together during the auto-zero cycle, they are also connected to the main op-amp's inverting input. Hence, nulling the input-offset voltage while common-mode voltage is present results in an extremely high CMRR. Power supply variations are nulled in a similar manner, and that's what gives the excellent PSRR mentioned above.

During the sampling cycle, the output of the first stage of the main amplifier is summed with the sampling circuit signal and then applied to the input of the second stage of the main amplifier. As the frequency of the input signal increases, error signals develop as the result of heterodyning (mixing of the input signal and the sampling signal). That produces error signals at a frequency f_E equal to the sum of or the difference between the sampling frequency, f_S, and the input frequency, f_I.

$$f_E = f_S \pm f_I$$

The *summed* error signal is of little concern because it is a high frequency that can be filtered out easily. However, when the input frequency approaches the sampling frequency (typically 330 Hz) the *difference* frequency approaches zero and that can cause dc errors—the problem that a chopper-stabilized amplifier is supposed to eliminate.

The solution is to filter the input to the sampling loop so that the circuit never sees any frequency that is near the sampling frequency. At a frequency well below the sampling frequency, the circuit forces the current out of the first stage of the null amplifier to equal the current through C1. The difference between them is zero, so the gain of the sampling loop is zero at that frequency, and at higher ones. Hence we have a low-pass filter whose cutoff frequency is determined by components in the output stage.

Each op-amp has a separate high-frequency amplifier; the −3-dB turnover frequencies for both the dc and the high-frequency paths are identical and equal. Therefore overall frequency response is smooth and free from sampling noise.

Practical Considerations

Circuit-board layout and the choice of external components both require careful consideration if low offset drift and bias current are to be maintained. A 20-page data application note has three pages of detailed applications information for the LTC1052 / LTC7652. Included is information that explains how to avoid the harmful effects to thermoelectric junctions, temperature transients, circuit-board leakage, and leakage in the holding capacitors.

Also included are eighteen practical circuit applications including a precision instrumental amplifier, a thermocouple amplifier with cold-junction compensation, a thermocouple-to-frequency converter, a 16-bit A-D converter and a voltage-to-frequency converter.

Precision Op-Amps

Four new high-reliability precision op-amps, processed to MIL-STD-883B specifications, have been announced by Solitron Devices. They feature high speed, low noise, low offset voltage and low drift. The OP-01 is a high-speed op-amp that features an 18 volt / μs slew rate, and a 0.7 μs switching time. The OP-07 is an ultra-low-offset-voltage op-amp that features 10 μv offset voltage, and drift of only 0.2 μv / °C. The OP-01 is available in a dual version (the OP-201), as is the OP-07 (the OP-207). The OP-01 and OP-07 conform to 741 pin-outs.

The op-amps are available in a variety of packages, including chip carriers. Military versions are available in hermetic metal cans as well as ceramic mini-DIP's and chip carriers.

Low On-Resistance Transistors

The 2N7010 and 2N7011 *FETlington* devices (JEDEC-registered by Siliconix) offer greatly reduced on-resistance and increased power dissipation. The newest members of the *FETlington* family of low-cost MOSPOWER products are designed to replace Darlington configurations with one compact DMOS device. The miniature TO-237 package offers a low-cost alternative to the larger TO-220 package, without changing the circuit-board design.

On-resistance of the devices is 0.35 ohm and they have breakdown voltages of 60 and 40 volts, respectively. They can handle continuous drain current of 1.3 amps. Maximum continuous power dissipation is one watt in free air, but dissipation doubles when the device is soldered to a conventional printed-circuit board.

Dual-Condition Sensing

Often we need to monitor two different conditions and raise an alarm when either condition falls above or below a normal value. Those conditions might be, for example, the outputs of a dual-voltage power supply, indoor and outdoor temperatures, or the high and low levels of the contents of a tank.

The electronics of such a two-input monitoring circuit has been greatly simplified by the introduction of the 3041 Monitor / Alarm IC by the Microcircuits Division of Intech. The 3041 is a dual-input circuit designed to monitor two different voltages and give an indication if either varies from the preset level by more than a user-settable predetermined percentage. The IC can provide a steady or an oscillating output, depending on the state of pin 4. The steady-state output is suitable for driving an LED, a lamp, or a TTL gate; the oscillating output is suitable for driving a speaker or a lightbulb.

The 3041 is packaged in an eight-pin mini-DIP. As shown in Fig. 3-11, it consists of a dual-input voltage comparator, a voltage regulator, and an oscillator. The 2.5-volt reference that is supplied by the regulator may vary between 2.2- and 2-7-volts dc. The reference voltage is available at pin 3. Maximum current drain (with the alarm on) is 13 mA from a 5-volt sup-

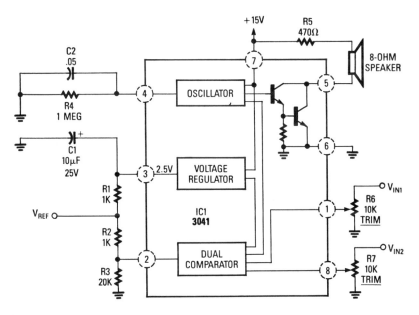

Fig. 3-11.

ply, and 25 mA from a 15-volt supply. Temperature stability is typically +50 ppm / °C.

The dual-input comparator monitors the voltages applied to pins 1 and 8, and compares them with reference voltage V_{REF}, which is present at the junction of external resistors R1 and R2. Actually, the voltage at pin 2 of the IC is used as the reference voltage, but the voltage at the junction of R1 and R2 is a more accurate representation of the voltage to which the comparator inputs are compared.

Only six discrete components are needed to build a basic monitor / alarm circuit. External resistors R1 and R2 are typically 1000 ohms each. Resistor R3 is determined by the desired "trip" tolerance, the percentage deviation from normal that will cause the alarm to trip. The resistor's value can be calculated as follows:

$$R3 = (100,000 \div \% \text{ Tolerance}) - 1000$$

The accuracies of resistors R1, R2, and R3 affect the tolerance of the comparator outputs by a factor of 10:1. For example, if the value of R3 is off by 5%, the comparator output will be off by 0.5%.

Capacitor C1 filters the voltage regulator's output, reduces noise to the comparator, and ensures a high-resolution trip point.

The internal oscillator can be disabled by grounding pin 4. For an ac output, connect an RC network to that pin as shown in Fig. 3-11. The values shown there will produce an output of about 1000 Hz; other frequencies are easily obtainable by using other components. Intech's data sheet contains a nomograph that aids component selection for various frequencies.

The 3041's open collector output (pin 1) can supply as much as 30 mA of current. The output is off (high) when both comparator inputs are within tolerance, and on (low) when either input is out of tolerance. With the component values shown in Fig. 3-11, when either V_{IN1} or V_{IN2} varies more than 5% from V_{REF}, a 1-kHz tone will be heard.

The 3041 can be used to monitor negative voltages. To do so, just connect pin 6 to the most negative point rather than to ground. If only one input is used, the other must be connected to the V_{REF} pin.

Ripple Detection

The 3041 can be used to detect ripple or oscillation at the inputs. To do that, connect a capacitor between the arm and the upper end of one of the input potentiometers. The dc voltage level is set by adjusting the pot while the ac signal is fed directly to the input through the capacitor. The circuit will respond to a peak ac voltage of 150 mV with a 5% tolerance setting, 230 mV with a 10% setting, and 420 mV with a 20% tolerance setting.

The value of the capacitor is determined by the value of the pot and by the lowest frequency of the signal that is to be detected. After setting the potentiometer for the desired dc level, measure the resistance between the arm and the high end. Select a capacitor whose reactance at the lowest frequency is small compared with the resistance it will be shunting.

Dual JFET Op-Amp

NEC Electronics has introduced the μPC812, a low-offset-voltage dual-JFET-input op-amp designed for stable operation when driving high-capacitance loads. Even when working into a 6800-pF load, the μPC812's output is absolutely stable. That stability is the result of using NEC's high-speed (f_T = 300 MHz) pnp transistor in the output stage. The high-speed transistor eliminates oscillation problems when sinking current into a large capacitive load.

Other features of the device include low input-offset voltage (3 mV maximum) and low input-offset voltage temperature drift (7 μV / °C). In addition, the device has output short-circuit protection, internal frequency compensation, and a high slew rate (13 V / μs).

Applications include fast sample-and-hold, high-speed integrators, and any circuit requiring high-performance input and output characteristics. The μPC812 is available in standard 8-pin plastic DIP and surface mount packages.

Quad Multiplying D/A

Precision Monolithic's DAC-8408 quad 8-bit, multiplying CMOS D / A converter with memory is designed for applications that require data-path verification, self diagnostics, and PC-board space savings.

The IC contains four identical 8-bit DAC's on a single CMOS chip. Each DAC has its own reference input, feedback resistor, data latches, and internal 3-state buffers. The IC uses an 8-bit TTL / CMOS-compatible data port that is bus-compatible with most 8-bit microprocessors, including the 6800, the 8080, the 8085, and the Z80. The port is common to all DAC's in the IC; doing so reduces pin count and allows the device to be packaged in a 28-pin, 0.6″ DIP.

The DAC-8408 features a read / write capability that allows each DAC's data word to be read from an addressable memory location. By reading the latched register contents back, the controlling computer can execute data-path verification, self-checks, and analog output updating.

The device is available in two grades in both military / industrial and commercial temperature ranges.

P-Channel JFET's

The 2N5460-2N5465 from Siliconix are low-cost P-channel JFET devices designed for a broad range of applications, including servo amplifiers, analog switches, level shifters, and control loops. Featured are high breakdown voltages of 40 to 60 volts, low operating currents of 1 to 9 mA, and high gain (1000 to 6000 μmhos, minimum). Fast switching is made possible by the devices' low inter-electrode capacitance (typically 1 to 5 pF). The JFET's come in TO-92 packages.

High-Power Darlingtons

The ZTX600 is the latest addition to Ferranti's family of E-Line (TO-92) npn Darlington transistors. The new series of high-performance, medium-power devices handle collector currents of as much as 1 amp, and that makes them ideal for automotive and industrial applications. Pulsed currents (I_{CM}) of up to 4 amps, and gains (h_{FE}) up to 100,000 at continuous currents ranging from 500-1000 mA are attainable.

The ZTX600 series offers V_{CEO} ratings of 140 to 160 volts. The construction of the E-Line package permits high junction temperature operation (200°C) that is usually associated with metal-can devices, which can dissipate up to 1.5 watts.

The ZTX600 series is available with several lead configurations and on radial or surface-mount tape for auto-placement.

Four MOSFET's

Four ITT-type MOSFET transistors that complement its line of N- and P-channel devices have been announced by Ferranti Semiconductors. The devices, BS170P, BS107P and BS107PT (N-channel), and BS250P (P-channel) are numbered according to the Pro-Electron or European system. They provide low threshold, low leakage and fast switching at breakdown voltages between 60 and 200 volts. The BS107P and BS107PT can withstand lightning surges of 1500 volts.

Transformerless 5-Volt Regulator

An ac-to-dc converter and voltage regulator in an 8-pin DIP IC is a new offering from Maxim Integrated Products. The six devices in the MAX600 series can reduce the cost, simplify the design, and reduce the component count, size, and weight of 5-volt dc, ½-watt power supplies. To create a 5-volt, 100-mA regulated supply, all you must add is a single filter capacitor. In addition, by adding a current-limiting resistor and a capacitor, four devices in the MAX600 series can connect directly to a 117-Vac power line.

The MAX600 and the MAX610 connect directly to the ac power lines and provide a five-volt output using an internal full-wave rectifier. The MAX601 and MAX611 are similar devices with half-wave rectifiers, and the MAX602 and MAX612 convert 8-volts rms to 5 volts dc using full-wave rectifiers. The MAX600, MAX601, and MAX602 have 0 to +50°C temperature ranges, and the MAX610, MAX611, and MAX612 have 0 to +70° ranges.

Contained in the 8-pin DIP package is a half-wave rectifier, a 12.4- or 18.6-volt Zener-diode shunt regulator, and a bipolar series-pass regulator. The nominal output voltage of all devices is 5 volts dc $\pm 4\%$; the output of the MAX600, MAX602, MAX610, and MAX612 can be set to any desired value between 1.3- and 15.0 volts dc.

A block diagram of the MAX600 series is shown in Fig. 3-12. Open-drain pin OUV goes low during under- and over-voltage conditions. The under- and over-voltage thresholds are fixed at 4.65 and 5.4 volts, respectively. Those thresholds do not change even if the output voltage is changed via the V_{SET} terminal, explained below.

Output voltage is determined by the state of pin 4, V_{SET}. If pin 4 is grounded, the output voltage will be the preset 5-volts

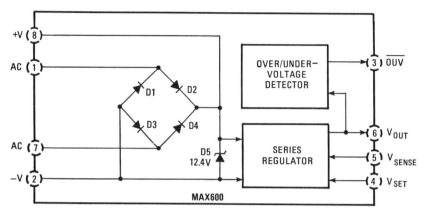

Fig. 3-12.

dc. Otherwise pin 4 can be used to set the output to any volt-age from 1.3 to 10 (for the MAX600/10) and from 1.3 to 15 (for the MAX602/12) by installing a simple resistive voltage divider. Pin 4 of the MAX601/11 controls a reset delay—the amount of time before pin 3 returns to a high level following an over- and under-voltage condition. The reset delay is directly proportional to the value of an external capacitor connected to pin 3. Each 0.01 μF of capacitance results in a 30-ms delay.

Pin 5 is the current-limit input. The output short-circuit current limit is 0.6V/R_{SENSE}, where R_{SENSE} is a current-limiting resistor connected between pins 5 and 6.

The rectified but unfiltered output of the diode bridge appears at pin 8. The desired filter capacitor should be connected between pins 8 and 2. The output of the regulator appears at pin 6.

Figure 3-13 shows a MAX600 configured as a 5-volt, 50-mA dc power supply. By substituting a 100Ω, 1-watt resistor for R1 and a 0.82-μF, 280-volt capacitor for C1, the circuit will run from a 220-volt, 60-Hz ac power line.

When output current demand is less than 10 mA, capac-itor C1 can be omitted; the available current will be determined by the value of R1. For 5-volt, 10-mA output, R1 should be 8200 ohms. Power dissipation is about 1.3 watts. For 220-volt ac operation, double the resistance and wattage.

NOTE WELL: The output of power supplies using a MAX600-series regulator is not isolated from the power line unless its input is supplied through an isolation transformer! The MAX600 device, its circuitry, and all components and

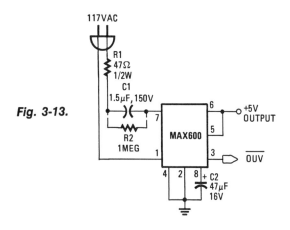

Fig. 3-13.

equipment driven by the 5-volt supply present a shock hazard and should be mounted in a protective enclosure to prevent accidental contact.

Further, when power is removed from a MAX600-based power supply, C1 may contain a charge equal to the peak value of the line voltage, thereby creating a second shock hazard. Therefore R2, the optional 1-megohm resistor, should be included to discharge C1 when power is removed from the circuit.

If the power supply is connected directly to the power line, do not connect the ground of an oscilloscope to the circuit. In addition to creating a shock hazard, doing so could severely damage a solid-state scope, as well as destroy the MAX600 device.

If the power supply must be isolated from the power line, you can use a 1:1 isolation transformer or a step-down transformer and a MAX602 or MAX612. The transformer should deliver 8-volts rms to maintain a regulated 5-volt output. The peak transformer output should not exceed 17 volts unless a series resistor is used to limit current to a safe value.

The maximum power dissipation of the MAX602 / 612 is approximately $(V_{IN} - V_{OUT}) \times I_{LOAD}$. With an 8-volt rms input, power dissipated in the device limits maximum output current to 100 mA at 25°C and 30 mA at 70°C.

If the 8-volt transformer is replaced by a 6.3-volt unit, maximum output current increases to 150 mA at 25°C, but the line voltage must not be permitted to drop below 100 volts. Otherwise, output voltage regulation will be lost. When using a 6.3-volt transformer, the capacitor connected to the +V terminal must be increased to 220 μF to help prevent +V from falling below 6.0 volts at any time.

Current-limiting capacitor C1 is critical when used in a 110 / 220-volt input supply. It should be non-polarized and rated for at least 150-volts rms. Metallized film capacitors are preferred to metal-foil types.

The value of C1 determines the power dissipated in the regulator IC and the maximum output current. It should be the smallest value that will deliver the desired output current at the minimum line voltage, because the power dissipated in the IC increases with the value of C1. For the full-wave MAX600 devices,

$$ C1 = \frac{I_{MAX}}{(V_{RMS} - V_{OUT}) \times 4\sqrt{2} \times F_{IN}} $$

where f_{IN} is the input line frequency. For half-wave MAX601 / 11 devices, the value of C1 is doubled.

Resistor R1 limits maximum peak current to 5 amps. That amount of current could flow if power were connected just as the instantaneous line voltage was at its maximum. With a 117-volt 60-Hz input, dissipation in mW is 1.6 × C1 × R1, where C1 is in microfarads and R1 is in ohms. For a 220-volt input, the constant in that equation is 2.7 instead of 1.6.

The maximum input to the MAX600 / 10 is 10 volts; those devices can supply outputs from 1.3 to 9 volts. Similarly, the maximum input to the MAX602 / 12 is 16 volts; those devices can supply outputs from 1.3 to 15 volts.

Temperature Transducer

The most commonly used temperature sensors and transducers include thermocouples, thermistors, temperature-dependent resistors, and biased diodes. The AD592 precision temperature transducer from Analog Devices provides output current that is proportional to absolute temperature; the IC is an ideal replacement or substitute for the types of transducers mentioned above. It is a two-terminal device that acts as a high-impedance temperature-dependent current source that yields 1 μA per degree Kelvin. It can operate from 4- to 30-volts dc.

The transducer can be used over a temperature range of $-25°C$ to $+150°C$ with typical calibration error of $2.0°C$ at $25°C$. Over the $0°C$ to $+70°C$ range, calibration error is $1.5°C$. Typical applications include temperature measurement and control in automotive, home, and industrial environments, HVAC (*Heating*, *Ventilating*, and *Air* Conditioning) system monitoring, and temperature correction in precision electronics. A low parts-count per application makes the AD592 a cost-effective device because expensive linearization circuitry, precision voltage references, and cold-junction compensation are not required.

The design and operation of the AD592 temperature transducer comes from basic silicon transistor theory: When two identical silicon transistors are operated so that there is a constant ratio between their collector currents, the differences in their base-emitter voltages will be directly proportional to absolute temperature.

In the AD592, the difference voltage is converted to a current that is proportional to absolute temperature by on-chip, low-temperature-coefficient thin-film resistors. The output current, when properly scaled, is equal to the absolute temperature (in degrees Kelvin) of the transducer.

During production, the on-IC scaling resistors are laser-trimmed to provide the 1 μFA / $°K$ output with supply voltages ranging between $+3$ and $+30$ volts. The output current ranges from 248 μA at $-25°C$, to 298 μA at $+25°C$, to 378 μA at $+105°C$.

The transducer is packaged in a plastic TO-92 case and is available in three performance grades with maximum calibration errors ranging from $0.5°C$ to $2.5°C$.

Figure 3-14 shows the basic circuit for using the AD592. Variable resistor R1 must be adjusted for the desired scale factor. To trim the circuit, a precisely known temperature must be measured by the transducer and the resistor adjusted for the desired output scale. For example, with the AD592 transducer at $0°C$, adjust R1 so that $V_{OUT} = 0$. Doing so nulls the initial calibration error and shifts the output units from Kelvin to Celsius.

By using an op-amp, scale factor and calibration errors can be eliminated; consult the manufacturer's data sheet for the circuit. Another circuit shown in the data sheet allows output to be scaled in either degrees Celsius or Fahrenheit. The circuit uses an op-amp, and allows for easier calibration.

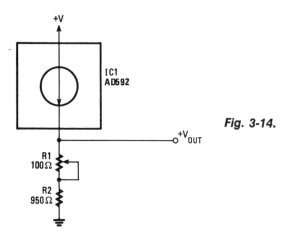

+V

IC1
AD592

+V_OUT

R1
100Ω

R2
950Ω

Fig. 3-14.

Bridge Motor Driver

Sprague's UDN-2998W, is a dual full-bridge motor driver that interfaces low-level logic to solenoids, brushless-dc and stepper motors. It can operate 50-volt inductive loads with continuous currents as high as two amps per bridge, and it can supply peak (start-up) current of as much as three amps per bridge. Control inputs are compatible with TTL, DTL, and five-volt CMOS logic.

The device differs from similar motor drivers in several respects:

- Eight power diodes (four per bridge) are included per IC. Those diodes, which are essential in motor-drive applications, are not provided on-IC by most other driver manufacturers.
- The 12-pin single-in-line power-tab package can dissipate 5.2 watts at +25°C—nearly 50% more than the 3.5 watts competitive devices can dissipate.
- An internal regulator allows operation from a single voltage power supply. Similar devices require an additional 5-volt supply.
- An internally generated turn-on delay prevents power-consuming, heat-producing crossover currents that would otherwise develop when switching phase (changing current direction).

Protection features include thermal-shutdown circuitry, crossover-current delays, and flyback and ground-clamp

116

diodes. For PWM (*Pulse Width Modulation*) control, an OUTPUT ENABLE pin is provided for each bridge. Sink-driver emitters pins are brought out for connection to external current-sensing resistors.

Quad Auto-Zero Op Amp

Teledyne Semiconductor's TSC914 is the first complete chopper-stabilized monolithic quad op-amp. Chopper-stabilized auto-zero amplifiers offer low offset voltage and drift by periodically sampling offset error, storing a correction voltage on a capacitor, and using that voltage to compensate for offset voltage and drift.

Earlier chopper-stabilized amplifiers have been packaged one to an IC, and earlier quad packages have been bipolar or low-performance CMOS types, neither of which can take advantage of chopper technology.

Unlike earlier chopper-stabilized op-amps, the TSC914's architecture makes possible the use of storage capacitors small enough to be included in the IC. The advanced chopper circuitry nulls offset voltage over time and with variations in temperature. Offset voltage is 15 μV maximum, drift is held to 0.15 μV / °C, and supply current is 850 μA maximum. Offset voltage is five times lower than the typical quad op-amp; offset voltage drift is eight times lower.

The TSC914's pinout matches that of the LM324. It is a drop-in replacement for the LM348, the OP-11, and the TL274, with $\pm$5-volt operation. The TSC914 is available in two performance and two package versions.

New Transistors

The 2N7056 and 2N7059, are the first in a series of transistors from Siliconix that utilize the company's MOSPOWER-6 technology in the electrically isolated TO-218 package. Those *ISOWATT218* devices provide 4000-volt isolation between internal electrical points and the heatsink or other mounting surface.

The devices are produced jointly by Siliconix and SGS Semiconductor and will be offered independently by both firms.

The 2N7059 has a 500-volt breakdown rating, maximum continuous current rating of 8 amps, and a maximum on resistance of 0.45 ohm. The 2N7056 is similarly rated at 200 volts,

19 amps, and 0.1 ohm. Each can dissipate 75 watts continuously without derating.

Long-time Timer

For most of the past decade, the mention of a monolithic timer immediately brought to mind the 555; the best-known and most widely produced timer IC. Although the 555 is a practical device for delay times ranging from a few milliseconds to several minutes, it becomes less reliable as the delay interval is increased. That's because the time interval is determined by an RC product, and long time intervals can require a very large-value capacitor, which usually means an electrolytic type.

But when accuracy is required you cannot use an electrolytic capacitor. For one thing, electrolytics are low precision devices; their value can drift with time. Finally, an electrolytic capacitor's inherent high leakage-current makes it impossible to use a high value of resistance in the circuit.

Though not as well known, a timer IC that is especially designed for long-time applications has been around since the early 1970's; it is the ZN1034E from Ferranti Electronics. When used as a stand-alone device, that IC can provide timed intervals ranging from 1 second to 19 days, although the RC time constant is only 220 seconds.

The ZN1034E includes an internal voltage regulator, an oscillator, and a 12-stage binary counter. The total delay time provided by the counter is 4095 times the period of the oscillator. Therefore, we can use moderate values of resistance and capacitance in the RC timing network and obtain periods that are many times longer than those possible with just the basic oscillator. With precision components with low temperature coefficients, the repeatability of timed periods is accurate to within 0.01% and the temperature drift in the timed period can be held to within 0.01% per degree Centigrade.

The ZN1034E comes in a 14-pin DIP package. A pinout and block diagram of the IC are shown in Fig. 3-15. Figure 3-16 shows a basic circuit for the device that is suitable for practical experimentation. Note that when you use a 5-volt power supply, only pin 4 is tied to the positive rail; pin 5 should not be connected. For supplies of from 6-450 volts dc, pins 4 and 5 are tied to the positive rail through dropping resistor R_D, as shown. The internal 5-volt regulator connected to pin 5 ensures that +5 volts is supplied to the internal circuitry.

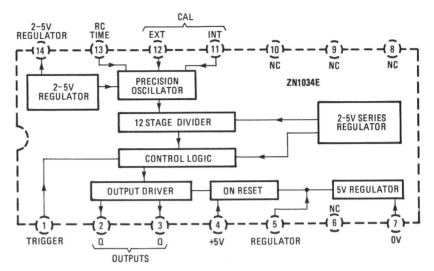

Fig. 3-15.

Resistor R_D must drop a voltage equal to V_{SUPPLY} minus 5 volts. Its value is derived from $(V_{SUPPLY} - 5) \div I$, where I is the load current plus the current drawn from either of the outputs.

The control logic times-out after 4095 cycles of the oscillator and delivers high and low output pulses at pins 2 and 3. The output at pin 3 is normally high and goes low at the end of the timed interval. The complementary output at pin 2 is normally low and goes high at the end of the timed interval. The timing period, in seconds, is obtained from the formula:

$$T = KR_T C_T$$

where R_T and C_T are the values of the resistor and capacitor respectively, and K is a multiplying factor determined by the resistance connected across pins 11 and 12. When pins 11 and 12 are connected together only an internal resistance of 100k is across those pins and the value of K is 2736. Connecting a 50k, 150k, or 300k resistor across those pins will provide multiplication factors of 2500, 4100, or 7500 respectively.

Resistor R_T can range from 5k to 5 megohms, but its value should be kept between 50k and 1 megohm for best performance and linearity. For best performance the value of C_T should be greater than 0.01 μF, but smaller values can be

Table 3-1.

R_T and C_T Values			
Timing Elements		**Timed Period**	
R_T (ohms)	C_T (μF)	**A**	**B**
39k	0.01	1 sec	2.92 sec
220k	0.1	1 min	2.75 min
100k	1.0	5 min	12.5 min
1.2 Meg	1.0	55 min	2.5 hrs
1.2 Meg	10	9.1 hrs	25 hrs
3.3 Meg	10	1 day	2.8 days
2.2 Meg	100	1 week	19 days

A) Pins 11 and 12 tied together
B) 300k resistor connected between pins 11 and 12

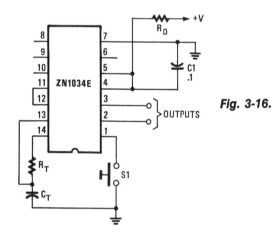

Fig. 3-16.

used when the timed interval must be short. With values below 0.01 μF, the timed interval is not linear with respect to the RC time constant. The manufacturer suggests 3900 pF as the minimum value for C_T. Table 3-1 shows convenient R_T and C_T values for timed periods ranging from 1 second to 19 days.

The timing period is initiated by momentarily grounding pin 1, in Fig. 3-16, that is done using a momentary pushbutton switch, S1. Alternately, if pin 1 is grounded, the timing period can be initiated by applying power to the circuit.

Figure 3-17 shows how the ZN1034E can be used as an interval timer providing delays of 1 to 11 minutes. Timing resistor R_T consists of two resistors, R1 and R2, in series.

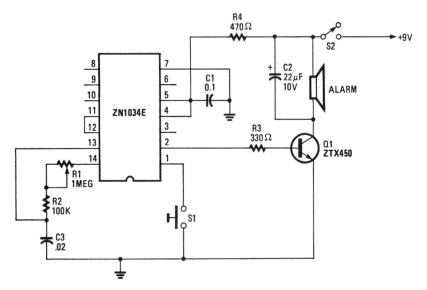

Fig. 3-17.

Because R1 is a fixed value of 100k, the total range of R_T is 100k to 1.1 megohms. Timing starts when S1 is pressed. The alarm sounds at the end of the timed interval and continues until S2 is opened. Transistor, Q2, a Ferranti ZTX450, is an npn medium-power (1 watt) silicon transistor with a V_{CBO} of 60 volts and an $I_{C(max)}$ of 1 ampere.

The ZN1034E's data sheet has circuits for a number of other applications, gives interfacing ideas, and shows how to cascade two timers to provide delays up to 1 year with an accuracy of 6 minutes. (Such long-timed periods can be used for time-lapse photography and for controlling automatic watering systems for lawns and greenhouses.)

SECTION 4

DESIGNER'S NOTEBOOK

Low-voltage Amplifier Circuits

DIGITAL AND ANALOG CIRCUITS each have their own unique set of design problems. Very often what is a major consideration in one field doesn't even appear in the other. There is, however, one problem that is common to both analog and digital circuits: the problem of tailoring real world signals so that they can be handled by whatever circuitry that is being designed to follow them.

The output of many real-world sensors (from microphones to keyboards to transducers) need a certain amount of conditioning before they can be reliably processed by either analog or digital circuitry. One of the most frequent problems that turns up is that the voltage level coming out of the input device is just too low to be used by the following circuitry. Because of that, those signals must be amplified to a usable level. We'll look at two general-purpose amplifier circuits. The first uses a single transistor and the other a CMOS IC. Either one can really come in handy when you're faced with the problem of low-voltage input signals. We'll look at the transistor amplifier first.

Transistor Amplifier

Figure 4-1 shows a simple single-transistor amplifier that can be used anytime when you need a boost for a signal that's

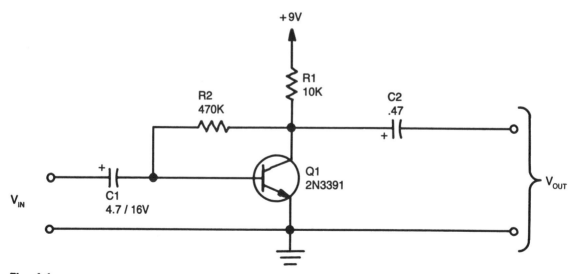

Fig. 4-1.

in the microvolt range. That circuit can be assembled from the sort of spare parts that fall into the cracks of your parts box. Not only that, but it uses so few parts that it takes up less space on a PC-board than an IC. In addition, the circuit has a flat frequency response across the audio spectrum and a gain of about 100 with the component values shown. None of the component values are particularly critical, therefore a wide range of substitutions can be made without seriously affecting the performance of the circuit.

The gain of the circuit can be lowered by dropping the value of feedback resistor R2. And the capacitor values shown can be changed if you don't happen to have those values on hand. Transistor Q1 is a small-signal high-gain npn transistor: substitutions can be made here as well. A 2N2222 transistor can be used but it will give you a lower gain than the 2N3391 shown; again it's a matter of trial and error on one hand, and how much gain you need on the other.

The circuit can be used anytime that a really low input signal needs to be boosted to a workable level. Anyone who has ever had to deal with the output level of a dynamic microphone (in the microvolt range) will find that little amplifier really handy because it will boost the mike's output signal level enough so that it can be fed into a standard line input. The other low-voltage amplifier circuit that we will look at uses CMOS inverters rather than a single transistor.

CMOS Amplifier

The second amplifier circuit is shown in Fig. 4-2. It uses three sections of a CMOS 4049 hex inverter IC (but any CMOS

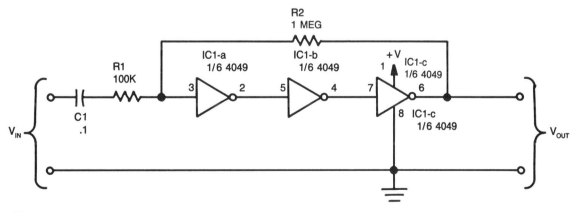

Fig. 4-2.

inverter can be used). It features a high input impedance. It also features all the good things we've come to expect from CMOS: a wide power-supply range, high noise-immunity, an output that swings from ground to just about the supply rail, and so on. The response of the circuit can be easily tailored to satisfy a wide range of circuit conditions.

The gain of that circuit is determined solely by the ratio of the feedback resistor (R2) to the input resistor (R1). And the frequency response is a function of the input capacitor. Keep this circuit in mind, it can make life a lot easier when the output signal for the circuit you're designing needs a bit of amplification. Just round up three spare inverters and your problem is solved. The only thing to remember is that the output voltage won't be at ground potential when you remove the input signal. Since we're using the inverters in a linear mode, the output voltage will always return to $V/2$ (where V is the supply voltage). If that presents a problem you can always take care of it with a capacitor or some other scheme at the output.

Charging Indicators

The attractiveness of nickel-cadmium or NiCd batteries for use as power sources, has caused them to begin showing up in many electronic parts catalogs and advertisements. Because they're still relatively expensive when compared with alkaline units, the advertisers in the back electronics publications offer them at a considerable discount; usually at least 30% off the regular price. And though nickel-cadmium batteries don't have the staying power of alkaline cells, being able to recharge them several times makes them extremely attractive for use in battery-powered devices. Also, they provide a constant voltage over the life of the charge, are relatively trouble free, and the charging circuitry for them is easy to design. All in all, if you take good care of them, they'll take good care of you.

However, one constant source of irritation in using NiCd cells has, paradoxlcally, nothlng to do with the batteries. The problem that we're referring to has to do with the charger, or more specifically the charging indicator. The irritation comes from the fact that the indicators are often misleading—they show that the charger is plugged in, but tell you nothing as to whether or not the batteries are actually taking a charge. And that's not all: There's a second source of trouble as well.

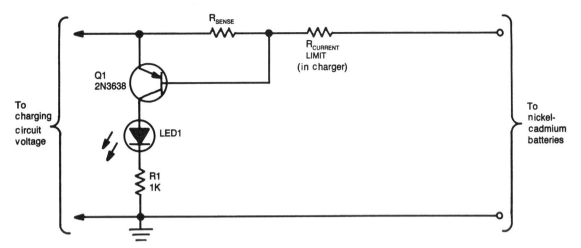

Fig. 4-3.

Let's assume that your charging circuitry doesn't include some sort of automatic changeover to trickle charge after the batteries have reached a certain charge level. If you keep pumping current into them at the same rate, you stand a good chance of blowing them up. Though NiCd cells may be available at discount prices, they're still not exactly cheap enough to destroy.

This circuit, shown in Fig. 4-3, is deceptively simple—it only calls for a handful of parts. But believe me when I say that it can save you a whole bunch of time, trouble, and most important, money. It gives you a way to make sure that the batteries are really charging and also tells you when they're fully charged.

How It Works

In the schematic shown in Fig. 4-3, transistor Q1 has its base-emitter junction connected across the sensing resistor (R_{SENSE}) on the line carrying the charging current. (Note that $R_{CURRENT\ LIMIT}$ is a part of the charger itself.) When the batteries are put on charge, current flows through the sensing resistor causing a voltage drop to be developed across it, and the resulting voltage turns on the transistor. With the transistor turned on, current flow through it causes the LED to turn on. However, the LED won't light unless the batteries are taking a charge! Sounds simple doesn't it?

Another feature of the circuit is that if the right value is chosen for the sensing resistor, the LED will extinguish when the batteries are fully charged, because of a charge in current flow through the circuit. Now, if the LED were part of some optoisolator arrangement, you could automatically increase the charger's current limiting resistor and cut the charge down to a trickle. Not bad for a handful of parts—and cheap ones at that!

We can't give you a value for the sensing resistor, because that depends on the amount of current needed to charge the batteries. However, *calculating* the resistance value needed is a piece of cake. Because we're using a silicon pnp transistor, it's going to take a voltage drop of about .65 volt to turn it on. The next thing you'll need to know is the charge rate of your unit. (Many chargers have their charge current and voltage printed on the wall transformer.) Once you have that information, the arithmetic is simple. The correct value for the sensing resistor can be found through the simple application of Ohms' law:

$$E = IR$$
$$R_{SENSE} = .65V / I_{CHARGE}$$

The value needed will typically be between 60 and 200 ohms.

Since the current-limiting resistor is usually much larger than 200 ohms, you can ignore the current limiting that the sensing resistor does. But try to keep that value as close to the calculated value as possible, because you want the transistor to turn off when the charging current starts to drop. If you're only interested in making sure that the batteries are really charging, you can forget the sensing resistor and put the transistor right across the current-limiting resistor. The parts for the circuit should cost you less than 50 cents and considering the price of NiCd batteries, that's a really cheap insurance policy!

Designing with the Schmitt Trigger

The Schmitt trigger is an incredibly useful circuit element whose popularity has increased since the introduction of the integrated circuit (IC). You can make a Schmitt trigger out of discrete components, but it requires that careful attention be paid to the math, plus the frequent use of non-standard component values. And even then, the results of hours of paperwork have a nasty habit of "blowing-up" when they're translated into real-world circuitry.

However, integrated circuits (and the control possible in manufacturing them) have changed all that. Today, Schmitt triggers show up in many applications, and, because of their unique properties, have simplified what used to be some really hairy circuit problems. We'll look at just one of the ways those circuits can eliminate all kinds of design hassles.

Schmitt-Trigger Oscillators

Everybody has seen the circuit in Fig. 4-4. If you have one Schmitt trigger, a resistor, and a capacitor, your clocking problems are solved. Any time you need an oscillator for some down-and-dirty clocking, you couldn't ask for anything simpler. Its output is a typical example of why Schmitt trigger IC's are such useful design elements. The symmetry of the output duty-cycle is well within shouting distance of a perfect 50-50; power requirements are low; and the circuit oscillates pretty much between ground and the supply rail. The output frequency, however, will depend on several factors including the supply voltage, and the threshold voltages for the IC. For 10-volt operation, the output frequency, f, of that circuit will be close to:

$$f = 0.72(RC)$$

A much more useful application of the Schmitt trigger in building an oscillator is shown in Fig. 4-5. That circuit is a low-power VCO (Voltage Controlled Oscillator) that is easy to build

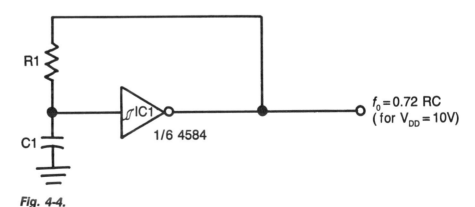

Fig. 4-4.

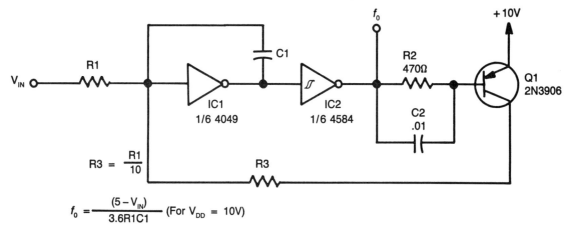

Fig. 4-5.

$$f_0 = \frac{(5 - V_{IN})}{3.6R1C1} \quad \text{(For } V_{DD} = 10V)$$

and has the added advantage of being linear over a wide range of voltages. The circuit is a bit more complicated, but much more versatile than the previous one. For openers, the output frequency can be calculated with a high degree of precision. For 10-volt operation, its output frequency is:

$$f = (5 - V_{in}) / (3.6(R1C1))$$

There are, however, some unusual things to watch out for when using it. For starters, the expression that shows how the frequency varies with the voltage is:

$$\frac{\Delta f}{\Delta V} = \frac{-1}{3.6(R1C1)}$$

Don't worry about where that expression came from—the important thing to note is that the minus sign indicates that as the voltage increases, the frequency decreases.

To understand what the limits of the circuit are, let's see how it works.

How It Works

The inverter (IC1, a 4049) acts as an integrator and produces a ramp voltage at its output. (A ramp is a linearly rising sawtooth wave, so named for its resemblance to an incline—like a staircase.) The integrator continues producing the ramp until its threshold voltage is reached. Because we're

130

talking about a CMOS inverter, the threshold voltage is equal to half the supply voltage. When the voltage at the input of IC2 reaches its threshold voltage, the output of the 4584 goes low and turns on transistor Q1.

Capacitor C2 begins to discharge rapidly, which produces a narrow negative-pulse at the output of the 4584. Capacitor C1 discharges as well, but the inherent hysteresis of the Schmitt trigger keeps the 4584 output low. Don't forget that the inverter changes state at half the supply voltage, and the Schmitt trigger changes at higher voltage than that. That's what hysteresis is all about.

In any event, the inverter finally changes states and the whole business starts over again. Now, the lower the input frequency, the faster both C1 and C2 will discharge. That means that as the voltage applied to the inverter decreases, the output frequency of the oscillator will increase, and vice versa.

The lowest frequency will be when the input voltage is zero and the maximum will be when the input voltage is equal to the threshold voltage of the Schmitt trigger. This is because the inverter and the 4584 will be changing states very close to each other. The circuit is extremely easy to use, and its advantages over other VCO's include a minimum parts count, high reliability, and, of course, excellent linearity.

Motor Speed Control

Small dc motors, like those usually found in toys, can be really handy things to keep around. They can be used in a variety of applications where the circuit you're designing has to move something other than just electrons. Those motors are great for control applications or just about anything else you can think of that doesn't require a great deal of precision.

However, those small dc motors have their own set of problems. For one, the motor's speed is notoriously dependent on the applied voltage. But that drawback can be turned into a benefit with the simple addition of a rheostat or potentiometer to make the motor speed variable. However, anyone who's ever tried making a motor-speed control using that principle, soon discovers that it's terribly unreliable, and it's an inefficient way to go about things.

No matter how small the motor, you still have to deal with the fact that they all have a certain amount of inertia, however small. That means that regulating really slow speeds with just a potentiometer is almost impossible. It usually requires that you get the motor going first and then back the potentiometer off until you achieve the desired speed.

Not only that, but if the motor draws a substantial amount of current, you're going to find that standard potentiometers won't be able to handle the power requirements: They'll start smoking and that will be the end of that. More expensive potentiometers can be used, but you'll still have the same low-speed problems. Obviously, there has to be a better way—and there is!

Controlling dc Motors

Another way of controlling the speed of a dc motor is shown in Fig. 4-6. Here instead of controlling the motor by varying the voltage, we simply apply a constant voltage and vary the duty cycle. All that means is that we'll control the amount of time the motor is on and allow the applied voltage to remain constant. By doing things that way, the inertia of the motor can be made to work on our behalf because it will keep the motor turning until the next pulse is applied to "kick" it along. Therefore, how fast the motor turns depends on the controlling oscillator.

Now, there are many ways to build an oscillator that can be used to control a motor. We've already seen that oscillator design is a wide open field and just about any combination of circuit building blocks can be used. Transistors, logic gates, 555 timers, and so on can be used to form the basis of a perfectly workable circuit. Each has its own advantages and disadvantages.

Figure 4-6 shows an oscillator circuit that may be used in motor-speed control applications: It is by no means the absolute last word in—or the best approach to—solving the problem. It is, however, one way to go about it and is perfectly workable in a wide variety of applications. In any event, that circuit will show you the basic method to follow in designing a circuit that is capable of handling your particular requirements.

Higher precision means using a more precise oscillator and adding a crystal to the circuit to lock-in the frequency. Heavier motors will need a "beefier" output stage than the single transistor shown in Fig. 4-6. However, that circuit shows

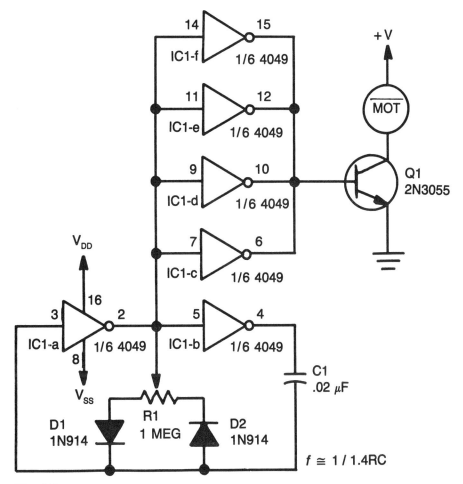

$$f \cong 1 / 1.4RC$$

Fig. 4-6.

the basic approach to follow (you may not find it necessary to go any further).

As shown, two inverters—IC1-a and IC1-b (each ⅙ of a 4048 hex inverter)—are used to make a simple oscillator whose frequency is approximately given by:

$$f = 1 / 1.4RC$$

Where R is the value of the potentiometer.

The basic circuit shown is one that you've seen a million times and have probably used just as often as a convenient clocking circuit. By adding the two diodes (D1 and D2), we can split the charging of the capacitor and thereby control the positive and negative parts of the cycle—D1 controls the positive and D2 controls the negative.

The time the circuit puts out a high and turns on the motor is controlled by the value of the left part of potentiometer R1 and the low-time is controlled by the right part of potentiometer. Regardless of how the potentiometer is set, the motor will always "see" the maximum voltage and as a result, the motor is less likely to stall at low speeds. You'll also find that the motor will start turning at a much slower speed than it would if the control was achieved by varying the voltage applied to it.

Although the transistor in our example is a 2N3055, any transistor that can handle the power requirements of the motor will do. If the motor is really "chunky," you might have to make a Darlington or add another stage to the output. Ganging the four remaining gates in the IC provides enough power to drive even a 2N3055, however, other applications may require other components. Although you can use any CMOS gate that can be made to oscillate, the 4049 is heftier than most of the others. But then, the final decision of circuit elements is yours, since only you know what your needs are.

There are several improvements on the basic circuit that come to mind almost immediately: They include using a crystal oscillator, adding a keyboard for speed-selection, or adding a digital display (which offers some interesting possibilities since all you have to do is count pulses and do a bit of arithmetic). Just as with all the circuits discussed in this column, remember that what we have described here is only a starting point.

A Simple Solution to Switch Debouncing

Some of the biggest headaches that show up in circuit design have absolutely nothing to do with electronics. That is, after you've spent all kinds of energy in taming electrons, the time comes when you have to connect the circuit to the outside world, and that's when the real trouble begins! Mechanical switching of electronic circuits is always an "iffy" business; and any designer who doesn't know that couldn't possibly recognize the symptoms, much less, solve the problem.

The most common causes of circuit "insanity" is what the data books refer to as "input-signal conditioning" or what the rest of the world calls debouncing. No mechanical switch is perfect, no matter how well it's made. As a result, pushing down on that little red button is going to generate more than one pulse. Any circuitry that's being triggered by that pulse is going to do exactly what it was designed to do—respond to each pulse it "sees."

There are all sorts of schemes to handle the problem. For one, you can use more expensive non-mechanical switches, or simply redesign the front-end of your circuit to respond to only one pulse. But the easiest way is to debounce the switch. There are dedicated IC's that can be used for that purpose but, as with most other things, there's an easier way.

Debouncing Circuits

The basic idea behind all switch debouncers is to put some type of isolating circuit between the switch and the circuit being triggered. The job of the extra circuit is to output one (and only one) pulse no matter how many bounces it "sees" from the switch. You can use anything from a flip-flop to a 555 timer (set up as a one-shot), but the problem can be handled a lot easier with inverters.

The most straightforward approach is to build a simple latch like the one in Fig. 4-7. Throwing switch S1 one way or the other will change the state of the output. Since the.e's

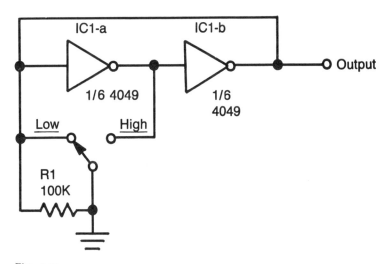

Fig. 4-7.

always some period of time during which no connection is made, resistor R1 is added to keep the circuit from glitching when the switch is thrown. That circuit is ideal for applications where you want to switch from one state to another. Even the noisiest single-pole, double-throw switch can be used because the resistor acts as a temporary storage device while the switch is being thrown.

The real problem appears when you want to use momentary (push-button) switches. That's because those switches are notoriously noisy, and if you don't take several precautions, they can screw up the operation of any circuit—no matter how well it's designed. Fortunately, there are two simple circuits that can take care of the problem.

The circuits in Fig. 4-8 are *half monostables* or edge detectors made from a single gate. The only difference between Fig. 4-8a and 4-8b are the gates: One is inverting and the other non-inverting. (We'll get to Fig. 4-8c in a moment.)

As you can see, the way the circuit responds depends on which end of the supply rail is tied to the resistor. The capacitor integrates the incoming switch bounces and causes the gate to change states. The capacitor then starts to discharge through resistor R1, and the gate (IC1) doesn't change back until its threshold voltage has been reached.

If you're still in the design stage of your circuit, you can add an extra IC to the board and get six (pushbutton) switch debouncers or three double-pole, double-throw debouncers. Doing the same thing with circuitry that's already in the circuit-board stage is a bit more difficult; you'll have to make a small "outrigger" board for the inverters. Remember that almost any inverting logic will do the job, so hunt around your design to see if you have any unused gates.

The values for the passive components depend on the type of switches you're using. In general, you should make sure that the output pulse is much longer than the switch bouncing. A good rule of thumb is to aim for a least a ten-to-one ratio. An output pulse width of 10 milliseconds should handle most bounce problems quite nicely. The values given in Figs. 4-8a and 4-8b should work for most applications. Just remember that the inverting gate will change the polarity of the input pulse, and the non-inverting one will preserve the polarity.

Although you can use any high-gain inverter to make a half monostable, the best all around choice is the Schmitt

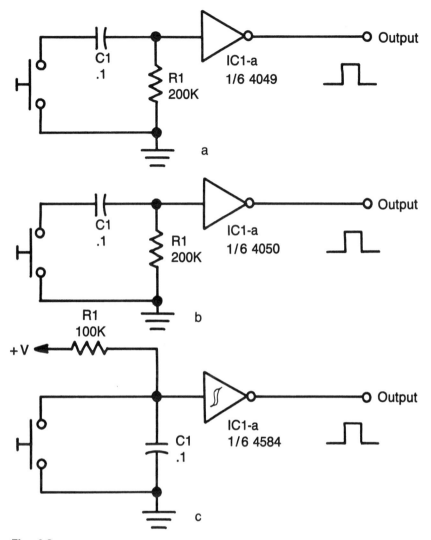

Fig. 4-8.

trigger. Not only do they have enough "zip" to respond properly, but its built-in hysteresis means you can get longer output
pulses.

Half-monostables use the capacitor as an integrator, but
you can also use it as a "sponge" to absorb extra pulses from
the switch, as shown in Fig. 4-8c. When the switch is open,
the input to the gate is held high, forcing the output of the IC
low. If there's bouncing when the switch is closed, the R-C
time constant keeps the glitching from affecting the output state
of the gate. Just as we saw earlier, though, make sure that

the time constant is going to be at least ten times the bounce time.

A lot of bench time has been wasted because pulses from a "bouncing" switch were masquerading as some other, more serious problem.

Audio Overload Protection

In any contest to rate the most popular areas of electronics, audio circuits and projects would undoubtedly be among the top ten. There is probably more home "tinkering" done in the areas of equalization, noise reduction, amplification, and so on than in any other field. And, as we all know, hardly a day goes by without an announcement from one semiconductor manufacturer or another about a new audio IC.

Each successive generation of audio IC has more features packed into it than its predecessor and can handle really mind boggling amounts of power. For instance, it wasn't long ago that an LM386 driver-amp blew everybody away because, with just a handful of external parts, it could output a ½ watt of continuous power into an 8-ohm load. These days, however, IC power-amps need virtually no external components, and one with more than 10 watts of power-handling capability can be held on the end of your little finger!

Every amplifier (regardless of type) has maximum power ratings. If those limits are exceeded, the amplifier and any associated components may be destroyed, so you must be careful. (Remember overloading can cause lots of trouble.) Overloading is hard to guard against because a typical audio signal can have a really wide dynamic range—sometimes more than 30dB.

Overload Protection Scheme

Protecting audio circuitry against overload (accidental or otherwise) is an important consideration, and should be on the mind of any serious audio-circuit designer. The best place to guard against overload is in the early stages where signal levels are low. The further along you are in the audio chain, the "beefier" the signal becomes, and the harder it is to add some type protection scheme. To complicate matters, overloads in the final power stages stand a much greater chance of "smoking" some expensive parts.

The circuit shown in Fig. 4-9 is the beginning of a protection scheme that can be made from a few common components. It's capable of monitoring circuit gain, and will also make sure that signal levels stay within the pre-set range. (The original circuit used a nonstandard optocoupler or optoisolator constructed from readily available parts, which we'll tell you how to make a little later.)

The best place to put the circuit is either in the feedback loop or shunted across the preamp input. Although the circuit tends to limit the gain of a preamp, keep in mind that it's meant to show you one way to approach the problem, and is by no means the only way to get the job done. Once you try it and become familiar with how it works, there are several "offshoots" of that design, which you can make following the same basic idea.

Figure 4-9 shows a 500-ohm potentiometer (R1) sitting right on the line feeding power to the preamp. When the audio signal is increased, the preamp draws more power to handle

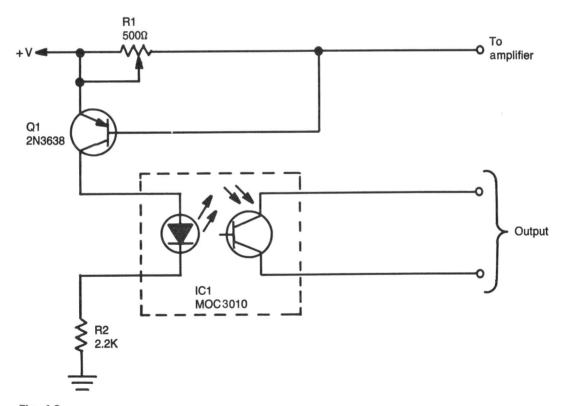

Fig. 4-9.

139

the larger signal. That results in a greater amount of current through R1, which causes a proportionate voltage to develop across the potentiometer.

Transistor Q1 monitors the voltage supplied to the amp through resistor R1. Whenever that voltage reaches the V_{CE} threshold, the transistor turns on, causing the LED in the optocoupler to light. That, in turn, causes the phototransistor to conduct. What you do with output of the optocoupler depends on how you design your audio circuit but (as already stated) the best place for it is either in the feedback loop or across the preamp input.

You can build an optocoupler using a *Light Dependant Resistor* (LDR), a jumbo red LED, and some heatshrink tubing. To make the optocoupler, simply place the LED and LDR inside the tubing so that the light from the LED can strike the lens of the LDR. Don't forget to allow the leads to extend beyond the tubing.

You can then use your optocoupler as you would any other pre-packaged type. Using that arrangement helps keep the active-component count to a minimum. The slow rise and fall times of the LDR actually work out for the best because it gives the action of the limiter a much more natural sound.

As I said, the general approach is more important to understand than the particular example. Figure 4-10 is an actual preamp using the familiar 741 as a non-inverting amplifier. As you see, the phototransistor (contained in the optocoupler) is connected in parallel with the feedback resistor. When an excessively high signal is pumped into the circuit, the amplifier draws more power to handle the increased input.

When the threshold voltage of the transistor is reached, it turns on the LED inside the optocoupler lights causing the phototransistor to turn on and lower the gain of the amplifier.

You could just as easily have connected the phototransistor or LDR from the input leg of the preamp to ground. However connecting it in such a manner requires a bit of recalculation of the resistor values in the circuit. Since I don't know what the change in voltage would be across your choice of optoisolator, you'll have to work the values out yourself.

Remember, because our approach to the problem of audio limiting is a general one, you'll have to tailor it to fit the specific needs of your circuit.

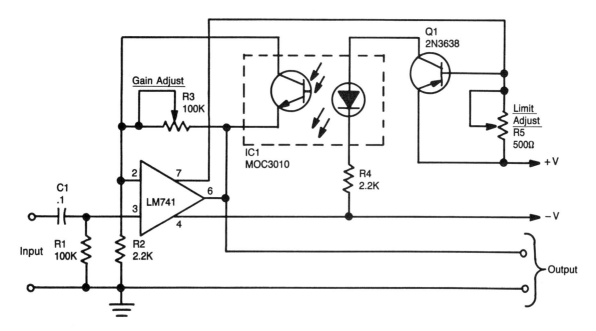

Fig. 4-10.

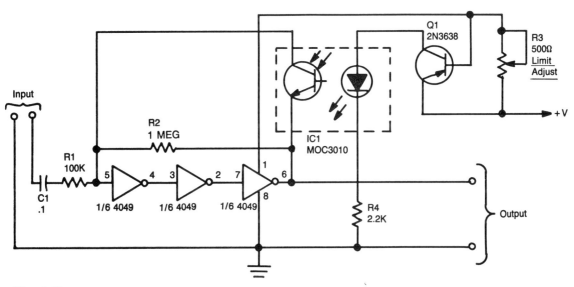

Fig. 4-11.

In Fig. 4-11, I've put the limiter to work in an amplifier made from a 4049 CMOS hex inverter. Since the gain of the circuit is only a function of R2 / R1, connecting the phototransistor in parallel with R2 will reduce amplifier gain whenever signal levels get excessive. The trigger for the circuit comes from the amp's current draw, rather than from the audio itself. That means that the gain is decreased before the amp overloads.

Our approach to limiting has several advantages over more conventional ones. It has a built-in failsafe because if anything happens to the LED, the phototransistor will not conduct (or the LDR will assume it's in-dark resistance—usually well over 1 megohm).

More On dc-Motor Speed Controls

A hard-and-fast rule in electronics is there's always another way to do something, and no matter how slick the design, it can always be improved. So, when I get an idea for some circuit, I usually try out the basic principle on a breadboard using whatever components I happen to have lying around. Once I'm sure the basic idea is sound, the task of improving on the design begins. And it's at that point that I usually find myself ordering components.

I presented a little circuit that could be used to control the speed of a dc motor. Since then, I've received several inquiries about it. The questions asked mostly dealt with the choice of components and a few details, which I guess weren't entirely clear in the article. Let me see if I can clear things up.

Controlling dc Motors

First of all, there was a typographical error in the schematic; the V + pin for the 4049 was incorrectly shown as pin 16 instead of pin 1. The 4049 is unique in the CMOS world because it has almost become a standard for the power pin to be directly across the IC from pin 1. The only saving grace is that pin 16 isn't used by the 4049, so the worst thing that can happen if you didn't catch the error is that the circuit won't work. (Usually, an error like that one would make things go up in smoke!)

The important part is making sure that you can design some way to control the duty cycle of the output. As with so many other things, it's not the particular circuit that's important, it's the idea behind the circuit. In other words, if you can understand the theory behind the circuit, you can design it any way you want.

To give you an idea of what I mean, let's look at Fig. 4-12. There we see a motor-speed controller that uses the most common oscillator I can think of—a 555.

If it looks at all confusing, cross out the diodes and you'll be looking at a simple 555 oscillator. The purpose of the diodes is to let us separate—and consequently control—the duration of the positive and negative halves of the output individually.

Although there are some minor practical differences between the operation of a plain 555 oscillator and the circuit shown in Fig. 4-12, you can use the standard design formulas to calculate the component values needed for your particular application. The frequency to aim for will depend on the motor you're trying to control. Generally speaking, the chunkier the motor, the lower the frequency you'll need. (Use that as a rule of thumb.)

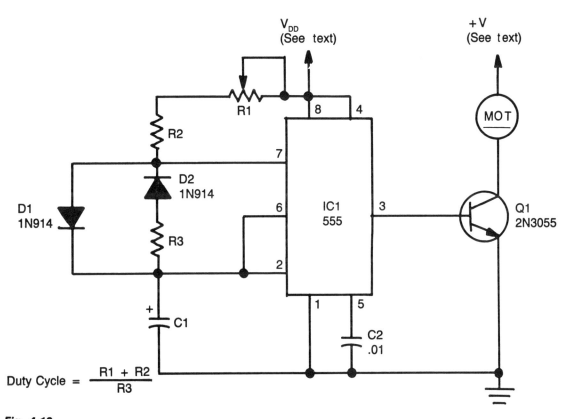

$$\text{Duty Cycle} = \frac{R1 + R2}{R3}$$

Fig. 4-12.

143

There are some advantages to using the 555 instead of the 4049 used originally. The most evident is that the 555 can provide 200 milliamps of drive current. Most small motors will whirl away happily with a somewhat smaller amount of current. But, if you're operating a heavier motor, you'll need to use the transistor output stage shown in the figure. If you have any doubts, put the transistor in the circuit.

If you're trying to control a motor that takes two men to lift, you can use the control circuit, but the design of the output stage is a whole separate matter. You'll need much more than a simple one-transistor switch. When you work out the values you'll need, remember that the duty cycle is controlled by the formula:

$$\text{Duty cycle} = R1 + R2 / R3$$

Also, several other readers have asked about the relationship between V_{DD} on the IC and V+ on the motor. Actually, they don't have any relationship. However, if the supply voltage for the motor falls within the allowable CMOS voltage range (3 to 15 volts), there's nothing stopping you from operating both the motor and the control circuit from the same supply.

The circumstances that first led to the development of the circuit that appeared involved a motor that ran on 48 volts . . . a bit outside the range of CMOS. In that case, I used a voltage divider to get the 12 volts needed to power the control circuit. But it would have worked just as well had I used a separate supply.

What you do is up to you. The motor-speed controller draws next to nothing in terms of power. However, if you do decide to supply V+ and V_{DD} separately, then what you'll have to do is make sure that they share a common ground. CMOS is marvelously noise immune, so things like motor noise has no effect on the circuit.

If you look the circuit over carefully, you should be able to figure out a way to not only control the speed of the motor, but the direction of rotation as well. If you understand what the circuit's doing, you can easily get the additional control for free. As a final note on this subject, remember that the circuit will power only dc motors. Controlling the speed of ac motors is something else altogether.

A Handy Low-voltage Indicator

I know it's hard to believe now, but once upon a time all electronics gear had to be powered from the ac lines. Batteries were put in flashlights, and nowhere else. Most other devices required much more power than a dry cell could deliver. These days, whenever I see someone walking down the street plugged into a tiny radio or cassette player, I think about companies like Webcor and get a bit nostalgic. Maybe some of you do too.

Anyway, more and more devices today use batteries for power, and that includes home-made devices as well. In the past we've presented several circuits that help keep batteries well-charged. We've also talked about nickel-cadmium cell charging and standby power supplies, but we've never shown how to keep an eye on batteries' energy reserves. And we all know (by Murphy's law) that batteries have a nasty habit of pooping out just when we need them most.

Well, I've used the little low-voltage indicator shown in Fig. 4-13 in almost every battery-powered device I ever built. The circuit has only five parts, and it pays for itself over and over in the amount of irritation it helps prevent.

How It Works

Input terminal V_{IN} is connected to the +V line of the circuit the indicator is to monitor, and the grounds of both circuits are connected together. The position of potentiometer R1's wiper determines Q1's base voltage. As long as the transistor gets enough bias to remain on, the low voltage at the collector will keep the SCR from firing. As the battery voltage starts to fall, the transistor's base voltage will fall as well.

The magic moment comes when the base voltage drops too low to keep Q1 on. When Q1 turns off, the collector voltage goes up, and that provides enough gate drive to turn on the SCR. That provides enough power to turn on the LED, which could also be buzzer or almost any other type warning device.

The circuit uses very little power. As a matter of fact, most current is drawn through the potentiometer, although some current flows through Q1 as long as it remains on. We've shown a value of 25 kilohms for R1, but you can use just about anything as long as you provide enough current to keep Q1 on. Of course, the larger the value of R1, the less wasted

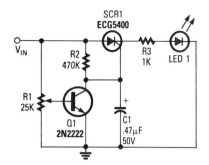

Fig. 4-13.

current there will be. Different 2N2222's will let you get away with larger potentiometers. As always, experiment. The capacitor is not strictly necessary; it's in the circuit to act as a "sponge" to prevent the SCR from firing when you change batteries.

The SCR

The last thing to discuss is the SCR. Remember that when the transistor turns off, the SCR is going to get its gate drive through the resistor. Not much current can flow through R2, so I've called for the most sensitive SCR I could find. If you need to drive something that draws more current than an LED, you'll have to adjust the back end of the circuit. There are many ways to do that: You could, for example, have the SCR trigger another SCR, or turn on a relay.

As it is, the circuit will work over a 25-volt range, and the potentiometer can be adjusted to make the circuit fire over most of that range. So, breadboard the circuit and try it out. I'm sure a bit of thought on your part will result in enhancements.

Schmitt Triggers

One of the main advantages digital has over analog electronics is the ease with which signals can be processed. After you've captured a signal, you can do just about anything you want with it. But it's often a problem in digital (and in analog) circuits to capture the signal in the first place.

The nice, logical circuit elements on the PC board want nice, clean input signals. And a clean input signal can be hard to come by. Often a signal from the "real world" must be conditioned before it can be handled by the electronics on your board. If the signal is dirty, it's a safe bet that even the world's best-designed circuit will take one look, turn up its electronic nose, and refuse to deal with it.

Probably the best, and simplest, way to condition a signal is to use a Schmitt trigger. All logic families (TTL, LSTTL, CMOS, HCMOS, etc.) have a collection of Schmitt trigger IC's in several different gate configurations. In Table 4-1 we list several popular, off-the-shelf devices. They're very easy to use, but sometimes you don't have room for another IC on a PC board, or cost considerations prevent you from adding another IC, and so on. So what you can do is build your own Schmitt trigger using an unused gate (and there's almost always one of those on a board).

Rolling Your Own

The circuit shown in Fig. 4-14 illustrates one way of converting a standard non-inverting gate into a Schmitt trigger. To understand how it works, assume that we're working with a CMOS gate, and that its input (and its output) are both low. What happens when we bring the input high? If R1 weren't connected to the input, the gate would change state a $\frac{1}{2} V_{cc}$, the normal CMOS trip point.

However, with the resistors connected, as the input begins to go positive, the output of the gate remains low. The resistors function as a voltage divider that keeps the voltage at the gate's input lower than the actual input voltage. The inverter will only change state when the voltage at its input is equal to half the supply voltage. The two resistors make that trip voltage greater than half the supply voltage.

When the gate finally changes state, its output goes high, and that output voltage will be higher than the gate's input voltage. When that happens, the voltage divider helps pull the

Fig. 4-14.

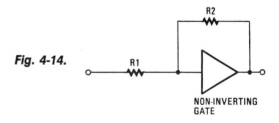

NON-INVERTING
GATE

gate's input high. So, if you think for a moment, you'll realize that what we've just described are the two basic characteristics of the Schmitt trigger—hysteresis and snap action.

The same sort of dynamics occur when the circuit goes the other way. The circuit trip point will be less than ½ V_{cc} by the amount of hysteresis generated by the resistors.

One advantage of designing your own Schmitt trigger is that you can play around with the values of the resistors and thereby create a Schmitt trigger with just as much hysteresis as you want. The range of hysteresis, H, you get is controlled by the supply voltage as well as by the values of the resistors:

$$H = (R1/R2)V_{cc}$$

There are limits to the values the resistors can have. You'll get no Schmitt-trigger action at all if R1's value is very small compared with R2's value, or if the resistors are inadvertently switched. A sensible limit for the values of the resistors is about 200k for R2, and a maximum R1:R2 ratio of 5:1.

Since the circuit is so easy to rig up, the best approach to finding the right resistor values for a particular application is probably the empirical one. Breadboard the circuit and then experiment with values for the two resistors. Just keep those values within the limits just mentioned. You could also create a chart of the amount of hysteresis produced by different combinations of resistors. Keep it handy so you can use it to build a custom Schmitt trigger whenever the need arises.

Precision Rectifiers

When you talk about rectifiers, the first thing that comes to mind is power supplies, but rectifiers are also used in many other circuits. Converting ac to dc is necessary in many RF circuits, and most circuits that measure real-world quantities first have to rectify sensor voltages. But even though regular diodes and bridges are adequate for many rectifying jobs, sometimes a different approach is needed.

The rectifying circuit you build for a power supply will work perfectly well when you're eliminating batteries, but it will be completely useless for RF. The reason is simply that the input voltage is less than the voltage needed to turn a diode on. Even small-signal germanium diodes require about 0.3 volts to turn on. That may not seem like much, but, if you are working with

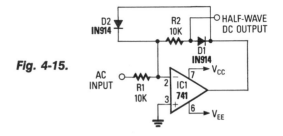

Fig. 4-15.

signals in the millivolt range, you'll have to find another way to handle the problem.

Circuit designers have two standard methods of dealing with the situation. They can amplify the ac signal and then rectify it, or they can do both at once with a precision rectifier. All things considered, the latter method is a much better way to get the job done.

The one-step approach to building a precision rectifier requires some way of isolating the positive and negative halves of the incoming ac, but after that ac has been amplified to a usable level. The circuit shown in Fig. 4-15 is a straightforward way of combining both amplification and rectification.

I designed the circuit with a 741 op-amp since it is cheap and readily available. If the performance specs of the 741 aren't to your liking, you can just as easily substitute any other op-amp. Higher input impedance, lower offset voltage, frequency limit, and slew rate are among the factors you should consider when choosing an op-amp. Examine the requirements of your application and choose an appropriate device.

How It Works

The circuit's theory of operation is similar to that of a diodes-only rectifier. During the negative half of the ac cycle the output of the 741 forward biases D2 and current flows only through that diode. During the positive half of the input swing, however, D1 is forward-biased, so current will flow through it and through R2. Therefore, dc will only show up across R2 during the positive part of the incoming ac cycle.

Because we're rectifying the voltage in the feedback loop of the op-amp and not at its input, the circuit will be able to handle very small ac signals. The inherent high gain of the op-amp allows us to rectify signals that are substantially below the voltage needed to forward-bias even small-signal germanium diodes.

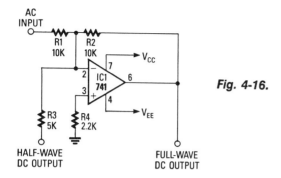

AC
INPUT

R1
10K

R2
10K

2

IC1
741

3

7

6

V_{CC}

V_{EE}

4

R3
5K

R4
2.2K

HALF-WAVE
DC OUTPUT

FULL-WAVE
DC OUTPUT

Fig. 4-16.

The op-amp shown in Fig. 4-15 is set up as an inverting amplifier, so the output waveform will be 180° out of phase with the input. You could switch inputs on the op amp to turn it into a non-inverting amplifier, but the phase difference comes in handy if you want to build a precision full-wave rectifier.

A simple summing amplifier can be used to turn our circuit into a precision fullwave rectifier, but a bit of thought has to go into picking the summing resistors. As shown in Fig. 4-16, we're adding the original ac signal and twice the output of the halfwave rectifier discussed above.

If R1 and R3 (in Fig. 4-16) had the same resistance, the output of the halfwave rectifier and the negative half of the input ac would be equal in magnitude, but 180° out of phase. In other words, the net result would be a voltage of zero. We can solve that problem by mixing in twice the halfwave voltage.

If you decide to build the fullwave rectifier, it's a good idea to use a 747, which has two 741's in a single IC package.

An Audio Oscillator

At various times I've presented you with several handy-dandy oscillator circuits of one kind or another. But if you're into building your own equipment you know the truth of Grossblatt's fifteenth law: *You can never have too many oscillators*.

An oscillator can be built from just about anything ranging from a handful of transistors to a rusty door hinge, but one circuit usually works better in a particular application than others. Some oscillators produce nice clean logic-level pulses; others are better suited for generating audio tones, etc. While reviewing past installments of this column, I realized that I've never shown you an oscillator that will drive an eight-ohm speaker. This circuit will correct that deficiency.

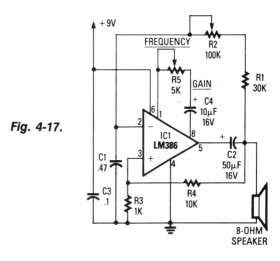

Fig. 4-17.

An audible tone generator is a useful circuit to have in the back of your drawer. It could be used as an alarm, a microphone tester, of any of a bunch of other things. For example, if you've ever done any recording that uses more than three or four microphones, you know how much trouble it is to keep the cables straight and to set uniform levels on the mixer. By attaching a battery-operated beeper to a microphone with a rubber band, the task will be much easier.

The oscillator shown in Fig. 4-17 is built around the popular and easily-obtainable LM386 low-voltage audio amplifier. Even Radio Shack carries it as one of their ever-dwindling stock of IC's. As you can see, the circuit requires only a handful of parts. The LM386 is such an even-tempered IC that it doesn't really care how you put those parts together. You could design a PC board to hold the parts, but the circuit will work just as well with wire-wrap construction.

The circuit's frequency of oscillation can be calculated easily from this formula: $f = 2.8 / [C1 \times (R1 + R2)]$. If you use the values I've indicated, you'll be able to vary the output from 60 Hz to 20 kHz by rotating potentiometer R2. There will be some droop in output at either end of R2's rotation, but, for microphone testing, that's not a problem.

How It Works

A portion of IC1's output voltage is fed to its noninverting input (pin 3). That voltage serves as a reference for capacitor C1, which is connected to the noninverting input (pin 2) of the IC. That capacitor continually charges and discharges around the reference voltage, and the result is a squarewave output. Capacitor C2 decouples the output so you get a nice ac signal to drive the speaker.

With GAIN control R5 at maximum, the circuit will deliver a fairly loud signal—about 40 dB according to my sound level meter. The LM386 has a 1350Ω resistor connected internally between pins 1 and 8. That resistor limits the minimum gain of the IC to about 20 dB. The 10-μF capacitor, C4, bypasses the resistor and allows you to vary the gain with an external control. If you don't need a gain control, just short pins 1 and 8 together.

Varying the size of output decoupling capacitor C2 also affects overall circuit gain. To change gain, you could experiment with its value, but you'll find that using a potentiometer and a capacitor across pin 8 is the easiest way to alter gain.

As with all circuits presented in this column, you can use them just as they appear here, but chances are that they'll be a lot more useful to you if you experiment with them to meet the needs of your specific application.

Trigger Pulses

See if this sounds familiar: You have put a circuit design on paper and have made a working breadboard. But as soon as you assemble the final version of the circuit the only thing it does is drain the battery. There are many reasons for that kind of problem, among them are wiring errors, bad printed-circuit traces, and intermittent components—and they should be the first things you check when things go wrong.

If those seem okay, the next thing you should check out is something that's often overlooked: *the state of the circuit when the power is first applied*. The more complex a circuit is, the more critical it is to make sure everything is set to a known state at power up. The parts needed to do that can be anything from a simple RC network to a separate active circuit designed to insure a specific start-up condition.

Most digital circuits need a pulse of predetermined length to set all the clocks, counters, latches, and so forth to a known state. The most basic way to generate a pulse is to use the circuit shown in Fig. 4-18, which consists of resistor R1, capacitor C1, and normally-open momentary-switch S1.

When the power is applied, capacitor C1 charges, and then discharges through resistor R1. As shown, the circuit

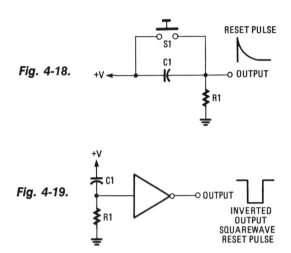

Fig. 4-18.

Fig. 4-19.

produces a positive-going pulse. A negative-going pulse is attained by simply reversing the power connections. The circuit is reset closing S1.

The simple RC pulse-generator shown in Fig. 4-18 will be adequate for designs that aren't particularly fussy about the shape or height of the reset pulse. Although the exact shape of the pulse will depend on the specific RC values, in some way it will resemble the shape shown in Fig. 4-18.

But suppose you need a pulse that resembles a clean squarewave with a wider width. The easiest way to generate such a pulse is to add a gate to the circuit of Fig. 4-18 and build a half-monostable; such a circuit is shown in Fig. 4-19. Although an inverter is shown in Fig. 4-19, you can use any kind of gate if you connect the inputs together. When the power is first applied to the half-monostable, it will change state and stay that way until the voltage at its input falls below the threshold voltage for the logic family. If you're using a CMOS device, that would be half of the power supply voltage. Then the half-monostable changes state again. The result is a clean pulse, whose width is determined by the value of R1 and C1. A close approximation of the pulse width is:

Pulse Width = .77RC

You can use almost any kind of gate (it can be inverting or noninverting) but Schmitt triggers are the best. They have considerable gain, and their hysteresis will guarantee a squarewave. A hex inverter is also a good choice for the circuit

Table 4-1.

| Schmitt Triggers | | |
IC	Type	Description
4093	CMOS	Quad 2-Input NAND Gate
4584	CMOS	Hex Inverter
40106	CMOS	Hex Inverter
74C14	CMOS	Hex Inverter
7413	TTL	Quad 2-Input NAND Gate
7414	TTL	Hex Inverter

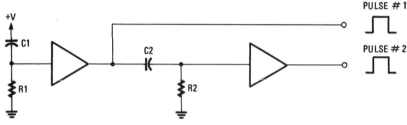

Fig. 4-20.

because it will make six half-monostables. That isn't as silly as it sounds because you can daisy-chain the individual circuits to get delayed pulses. Use the circuit shown in Fig. 4-20 if you need a series of pulses in a particular sequence and you have to be able to set different widths for each pulse.

Half-monostables are edge-triggered devices, so if you plan on daisy-chaining several circuits you'll have to pay attention to the polarity of the pulses. If necessary, you can use a small-signal transistor as a flip-flop to convert positive pulses to negative.

When you're working with a complex design, you should pay as much attention to the reset as you do to any other part of the circuit. Its timing can be very critical, and if it's off the mark you may get intermittent operation.

More complex reset problems require more complex solutions. You can find dedicated IC's that are designed specifically to generate pulses of one kind or another, but they are often expensive, hard to find, and difficult to use. Half-monostables are simple, cheap, and easy to use. The next time you find yourself looking for some way of generating a variety of pulses, use a hex inverter and wire up a couple of half-monostables. Chances are they'll do the job with a minimum of fuss and bother.

INDEX